Springer Theses

Recognizing Outstanding Ph.D. Research

Aims and Scope

The series "Springer Theses" brings together a selection of the very best Ph.D. theses from around the world and across the physical sciences. Nominated and endorsed by two recognized specialists, each published volume has been selected for its scientific excellence and the high impact of its contents for the pertinent field of research. For greater accessibility to non-specialists, the published versions include an extended introduction, as well as a foreword by the student's supervisor explaining the special relevance of the work for the field. As a whole, the series will provide a valuable resource both for newcomers to the research fields described, and for other scientists seeking detailed background information on special questions. Finally, it provides an accredited documentation of the valuable contributions made by today's younger generation of scientists.

Theses are accepted into the series by invited nomination only and must fulfill all of the following criteria

- They must be written in good English.
- The topic should fall within the confines of Chemistry, Physics, Earth Sciences, Engineering and related interdisciplinary fields such as Materials, Nanoscience, Chemical Engineering, Complex Systems and Biophysics.
- The work reported in the thesis must represent a significant scientific advance.
- If the thesis includes previously published material, permission to reproduce this must be gained from the respective copyright holder.
- They must have been examined and passed during the 12 months prior to nomination.
- Each thesis should include a foreword by the supervisor outlining the significance of its content.
- The theses should have a clearly defined structure including an introduction accessible to scientists not expert in that particular field.

More information about this series at http://www.springer.com/series/8790

Jing Guo

High Resolution Imaging, Spectroscopy and Nuclear Quantum Effects of Interfacial Water

Doctoral Thesis accepted by
Peking University, Beijing, China

Author
Dr. Jing Guo
School of Physics, International
  Center for Quantum Materials
Peking University
Beijing, China

Supervisor
Prof. Ying Jiang
School of Physics, International
  Center for Quantum Materials
Peking University
Beijing, China

ISSN 2190-5053				ISSN 2190-5061 (electronic)
Springer Theses
ISBN 978-981-13-1662-3			ISBN 978-981-13-1663-0 (eBook)
https://doi.org/10.1007/978-981-13-1663-0

Library of Congress Control Number: 2018948611

© Springer Nature Singapore Pte Ltd. 2018
This work is subject to copyright. All rights are reserved by the Publisher, whether the whole or part of the material is concerned, specifically the rights of translation, reprinting, reuse of illustrations, recitation, broadcasting, reproduction on microfilms or in any other physical way, and transmission or information storage and retrieval, electronic adaptation, computer software, or by similar or dissimilar methodology now known or hereafter developed.
The use of general descriptive names, registered names, trademarks, service marks, etc. in this publication does not imply, even in the absence of a specific statement, that such names are exempt from the relevant protective laws and regulations and therefore free for general use.
The publisher, the authors and the editors are safe to assume that the advice and information in this book are believed to be true and accurate at the date of publication. Neither the publisher nor the authors or the editors give a warranty, express or implied, with respect to the material contained herein or for any errors or omissions that may have been made. The publisher remains neutral with regard to jurisdictional claims in published maps and institutional affiliations.

This Springer imprint is published by the registered company Springer Nature Singapore Pte Ltd.
The registered company address is: 152 Beach Road, #21-01/04 Gateway East, Singapore 189721, Singapore

Supervisor's Foreword

For most of quantum materials, their properties are dictated by the quantum behavior of electrons, while the nuclei are only treated as classical particles. However, light nuclei like H^+ (proton) can exhibit prominent quantum effects due to the small mass, in terms of tunneling and zero-point motion. The so-called nuclear quantum effects (NQEs) are responsible for many abnormal properties of light-element materials such as water. Unfortunately, the accurate and quantitative description of NQEs is very challenging in experiments, since the quantum states of nuclei are much more fragile than those of electrons and are extremely sensitive to the atomic-scale environments. In this thesis, Dr. Jing Guo has developed ultrahigh-resolution scanning probe microscopy/spectroscopy, which allows the access to the quantum degree of freedom of protons, in addition to that of electrons, both in real and energy space. Combined with *ab initio* path integral molecular dynamics (PIMD) simulations, Dr. Guo and her collaborators have made groundbreaking steps toward understanding the NQEs of water/ice.

In her first work, Dr. Guo developed a novel submolecular imaging technique, which allows locating in real space the position of protons. Based on this technique, Dr. Guo succeeded to track the proton transfer in real time within a water nanocluster and observed quantum tunneling of multiple protons in a concerted fashion. It is striking that such a collective proton tunneling can readily occur even near the zero temperature.

This work puts an end to a 20-year-long debate whether the collective proton tunneling exists or not and provides an answer to the well-known puzzle: nonzero entropy of ice at zero temperature. The concerted proton tunneling in water/ice has been confirmed afterward by many groups (both theoretical and experimental) using different techniques.

Another fundamental question of NQEs is how the zero-point motion of proton affects the H-bond interaction. Dr. Guo provided a smoking gun for this question using a new technique called tip-enhanced inelastic electron tunneling spectroscopy (TEIETS). She and her collaborators unraveled quantitatively, for the first time, the quantum component of a single H bond at a water–solid interface and arrived at a general picture that the zero-point motion of protons weakens the weak H bonds but strengthens the relatively strong ones.

This work yields a cohesive picture for the quantum nature of H bonds. Those findings may provide answers to many weirdness of water from a quantum mechanical view. Furthermore, it is helpful for understanding exotic quantum behaviors of other H-bonded materials such as the high-temperature superconducting phase of H_2S.

Dr. Guo's creative approaches and her remarkable findings add substantially new insights into quantum matters. Those findings may completely revamp our understanding of water and other light-element materials from a full quantum perspective. Dr. Guo's work also provides new dimensions to manipulate the quantum properties and opens up a new frontier of condensed matter physics.

Beijing, China
June 2018

Prof. Ying Jiang

Preface

The mystery of water mainly arises from the intermolecular hydrogen-bonding (H-bonding) interaction. In this thesis, I will focus on the development of advanced high-resolution imaging and spectroscopy techniques based on scanning tunneling microscopy (STM) and noncontact atomic force microscopy (nc-AFM), which enable atomic-scale investigation of surface water, such as probing the O-H directionality and H-bonding interactions both in real and energy space, and exploring the quantum nature of protons, including proton tunneling and quantum fluctuations.

First, I report the use of STM to achieve submolecular-resolution imaging of individual water monomers and tetramers by employing the STM tip as a top gate to tune controllably the molecular density of states (DOS) of water around the Fermi level. I will also introduce the use of a qPlus-based nc-AFM to actualize the nearly noninvasive submolecular-resolution imaging of the water clusters by probing the high-order electrostatic force between the quadrupole-like CO-terminated tip and the polar water molecules at relatively large tip–water distances. These techniques allow us to discriminate in real space the O-H directionality and pin down the peculiar atomic-level structures of water monomers and clusters on NaCl(001) surface, which have not been reported before. Second, I introduce a tip-enhanced strategy to push the limit of vibrational spectroscopy of a prototypical hydrogen-bonded system down to the single-bond level by measuring the inelastic electron tunneling spectroscopy (IETS) in the near-resonance region. Comparing with the conventional IETS, the signal-to-noise ratios of the tip-enhanced IETS are enhanced by orders of magnitude, which provide a sensitive probe for NQEs of protons in energy space.

Third, I demonstrate the direct evidence of the concerted proton tunneling and also highlight the important role of individual ions in affecting the correlated tunneling process by monitoring in real time reversible interconversion of the H-bonding-associated chirality of the tetramer based on the submolecular-resolution imaging technique. Finally, I report the direct assessment of the quantum component of the H bond in a quantitative way and reveal the picture that the anharmonic quantum fluctuations of hydrogen nuclei weaken the weak hydrogen bonds and strengthen the strong ones through isotopic substitution experiments using the tip-enhanced IETS technique. It is particularly striking that the coupling to the atomic-scale species may completely reverse the widely accepted behavior of the NQEs, implying that the NQEs should be highly inhomogeneous and dynamic in nature, which is at present inaccessible by conventional spectroscopic methods.

Technically, the high-resolution imaging and spectroscopy techniques developed in this thesis open up the possibility of determining the detailed topology and intermolecular interaction of water networks and other hydrogen-bonded systems, such as ion hydration and biological water with atomic precision. Scientifically, the atomic-scale investigation of NQEs of surface water shed new light on the understanding of many weirdness and the macroscopic properties of water and ice from a quantum mechanical view.

Beijing, China
May 2018

Dr. Jing Guo

Part of this thesis has been published in the following journal articles:

1. J. B. Peng*, **J. Guo***, P. Hapala*, D. Y. Cao, R. Z. Ma, B. W. Cheng, L. M. Xu, M. Ondráček, P. Jelínek, E-G Wang, Y. Jiang, Weakly perturbative imaging of interfacial water with submolecular resolution by atomic force microscopy. *Nature Communications*. 9, 112 (2018).
2. **J. Guo***, J-T Lü*, Y. X. Feng*, J. Chen, J. B. Peng, X. Z. Meng, Z. C. Wang, Z. R. Lin, X-Z Li, E-G Wang, Y. Jiang, Nuclear quantum effects of hydrogen bonds probed by tip-enhanced inelastic electron tunneling. *Science*. 352, 321 (2016).
3. X. Z. Meng*, **J. Guo***, J. B. Peng*, J. Chen, Z. C. Wang, J-R Shi, X-Z Li, E-G Wang, Y. Jiang. Direct visualization of concerted proton tunneling in a water nanocluster. *Nature Physics*, 11, 235 (2015).
4. **J. Guo***, X. Z. Meng*, J. Chen*, J. B. Peng, J. M. Sheng, X-Z Li, L. M. Xu, J-R Shi, E-G Wang, Y. Jiang, Real-space imaging of interfacial water with submolecular resolution. *Nature Materials*. 13, 184 (2014).
5. J. Chen*, **J. Guo***, X. Z. Meng*, J. B. Peng, J. M. Sheng, L. M. Xu, Y. Jiang, X-Z Li, E-G Wang, An unconventional bilayer ice structure on a NaCl(001) film. *Nature Communications*. 5:4056 DOI: 10.1038/ncomms5056 (2014).

(* equal contributors)

Acknowledgements

The five years I am studying and living at Peking University is the most wonderful experience in my life. I deeply appreciate the help and strong support from my family, advisor, collaborators, colleagues, and friends.

First of all, I would like to thank my advisor, Prof. Ying Jiang, for his instruction and inspiration. At the beginning of my research work, Prof. Jiang spent a lot of time working with me to show how to use and manipulate the experimental equipment and how to carry out the experiment. It is his great patience and strong support that encourage me to step into the research world and keep on exploring the new things. I am always inspired and enlightened during the discussion with Prof. Jiang and deeply influenced by his passionate enthusiasm and great diligence to the scientific research. I also want to thank Prof. En-Ge Wang and Prof. Xinzheng Li for the fruitful discussion and suggestions, which is not only helpful to my research work, but also beneficial to the future research career. My sincere gratitude is also due to the other collaborators, Dr. Ji Chen, Dr. Yexin Feng, and Prof. Jingtao Lü, for their great support from the theoretical part. Without them, there is no possibility to achieve the works in this thesis.

I also would like to acknowledge my colleagues and other members that have studied in Prof. Jiang's laboratory. Dr. Xiangzhi Meng is very experienced in designing and building experimental instrument, and he has shared many valuable experiences about how to keep the instrument working in a good and stable performance. I admire his patient, careful, and selfless character. Dr. Jinbo Peng has done a lot in analyzing the experimental data of concerted proton tunneling in water tetramers. It is a great pleasure to work with Jinbo, because he is a very kind person and has endless curiosity about science. I also enjoy the working experience with Zhichang Wang because of his thorough understanding of physical picture and optimistic, open-minded personal character. I also want to thank Zeren Lin for her help for the fitting of the vibrational spectroscopy, Jiming Sheng for the LabVIEW programming, and Mingcheng Liang for the movie production for our *Science* work. I am also grateful to the other group members, Ke Bian, Runze Ma, Chaoyu Guo, Sifan You, Qin Wang, for the direct or indirect help and support.

I also want to thank Prof. Andrew Hodgson, who has invited me to visit Liverpool University. I deeply appreciate his warm help and guidance during the time I was in England. What's more, Prof. Andrew Hodgson has provided invaluable advice to our works.

Last, but the most important, I would like to thank my parents for their endless love and support. When I was a young girl, they told me "Good good study, day day up." I always keep these simple words in mind and try to make it. Finally, I want to express my great gratitude to my husband, Mr. Liang Ma, and my 21-month-old baby son. I love you all.

Contents

1 Introduction	1
1.1 Water-Solid Interface	2
1.1.1 Structure of Water on Metal Surfaces	2
1.1.2 Adsorption of Water on Metal Oxide Surfaces	5
1.1.3 Dynamics of Water on Solid Surfaces	6
1.2 Nuclear Quantum Effects of H-Bonded System	8
1.2.1 Nuclear Quantum Effects	8
1.2.2 NQEs of Water and Ice	8
1.2.3 NQEs in Other H-Bonded Materials and Biological Systems	11
1.3 Structure of This Thesis	12
References	14
2 Scanning Probe Microscopy	23
2.1 Scanning Tunneling Microscopy	24
2.1.1 Principle	24
2.1.2 Scanning Tunneling Spectroscopy	27
2.1.3 Inelastic Electron Tunneling Spectroscopy	29
2.1.4 Applications	32
2.2 Non-contact Atomic Force Microscopy	34
2.2.1 Principle	34
2.2.2 Q-plus Sensor Based nc-AFM	35
2.2.3 Applications	36
References	37
3 Submolecular-Resolution Imaging of Interfacial Water	43
3.1 Introduction	43
3.2 Methods	44
3.2.1 STM/AFM Experiments	44
3.2.2 DFT Calculations	45
3.2.3 AFM Simulations	46

3.3	\multicolumn{2}{l}{Submolecular-Resolution Imaging of Interfacial Water with STM}	47

3.3 Submolecular-Resolution Imaging of Interfacial Water
with STM 47
    3.3.1 Orbital Imaging of Water Monomers 47
    3.3.2 The Mechanism of Orbital Imaging 51
    3.3.3 Discrimination of H-Bond Directionality of Water
          Tetramers 53
    3.3.4 Characterization of H-Bonded Water Nanoclusters...... 56
    3.3.5 An Unconventional Bilayer Ice 58
3.4 Non-invasive Imaging of Interfacial Water with AFM 58
    3.4.1 Background 58
    3.4.2 AFM Images of Two Degenerate Water Tetramers
          with a CO-Terminated Tip 60
    3.4.3 The Role of Electrostatics in the High-Resolution
          AFM Imaging of Water Tetramers 61
    3.4.4 Submolecular-Resolution AFM Images of Weakly
          Bonded Water Clusters 65
    3.4.5 Quantitative Characterization of the Non-invasive
          AFM Imaging Technique 66
3.5 Summary 67
References 69

4 Single Molecule Vibrational Spectroscopy of Interfacial Water 73
4.1 Background 73
4.2 Resonantly Enhanced IETS 74
4.3 Methods 74
4.4 Tip-Enhanced IETS of Water Monomers 75
    4.4.1 Selective Orbital Gating of Water Monomers
          with Cl-Tip 75
    4.4.2 Single Molecule Vibrational Spectroscopy of Water
          Monomers 76
    4.4.3 Lineshape Change of Tip-Enhanced IETS 76
4.5 Action Spectroscopy 78
4.6 Summary 80
References 80

5 Concerted Proton Tunneling 83
5.1 Introduction 83
5.2 Chirality Switching of Water Tetramers 84
5.3 Quantitative Analysis of the Switching Rate 85
    5.3.1 Impact of Bias on Switching Rate 85
    5.3.2 Dependence of Switching Rate on Temperature 86
    5.3.3 Isotope Effect of Switching Rate 86
    5.3.4 Energy Profiles of Chirality Switching in Water
          Tetramers 87

	5.4	Impact of Local Environments on Concerted Proton Tunneling	89
	5.5	Summary	92
		References	92
6	**Nuclear Quantum Effect of Hydrogen Bonds**		**95**
	6.1	Introduction	95
	6.2	Measurement of H-Bonding Strength by Tip-Enhanced IETS	96
		6.2.1 Tip-Enhanced IETS of Stretching Mode	96
		6.2.2 Tuning of H-Bonding Strength	97
		6.2.3 Extraction of the Intrinsic Vibrational Energies	99
	6.3	Impact of NQEs on the Strength of a Single H Bond	99
		6.3.1 Tip-Enhanced IETS of HOD Monomers	99
		6.3.2 NQEs of H-Bonding Strength	101
	6.4	The Picture of Competing Quantum Effects	103
	6.5	Summary	105
		References	106
7	**Outlook**		**109**
	7.1	Perspective on Future Directions	109
		7.1.1 Overlayer Ice	109
		7.1.2 Confined Water	110
		7.1.3 Water Hydration	110
		7.1.4 Nuclear Quantum Effects	111
	7.2	Challenges and New Possibilities	112
		7.2.1 Ultrafast H-Bonding Dynamics	112
		7.2.2 Non-invasive Measurement of Nuclear Spin of Proton	112
		References	113

Abbreviations

AS	Anticlockwise state
C_6H_{12}	Cyclohexane
cNEB	Climbing image nudged elastic band
COF	Covalent organic framework
CS	Clockwise state
D	Deuterium
DFT	Density functional theory
DOS	Density of states
EELS	Electron energy loss spectroscopy
E_F	Fermi level
H bond	Hydrogen bond
H_3S	Hydrogen sulfide
HF	Hydrogen fluoride
HOMO	Highest occupied molecular orbital
IETS	Inelastic electron tunneling spectroscopy
IR spectroscopy	Infrared spectroscopy
LUMO	Lowest unoccupied molecular orbital
MC simulations	Monte Carlo simulations
MD simulations	Molecular dynamics simulations
MOF	Metal–organic frameworks
nc-AFM	Noncontact atomic force microscopy
NMR	Nuclear magnetic resonance
NQEs	Nuclear quantum effects
NV center	Nitrogen-vacancy center
PDOS	Projected density of states
PES	Potential energy surface
PIMD simulations	Path integral molecular dynamics simulations
SFG	Sum-frequency generation spectroscopy
STM	Scanning tunneling microscopy
STS	Scanning tunneling spectroscopy

TI	Thermodynamic integration
VASP	Vienna *ab initio* simulation package code
vdW	van der Waals
XPS	X-ray photoelectron spectroscopy
XRD	X-ray diffraction
ZPE	Zero-point energy
ZPM	Zero-point motion

Chapter 1
Introduction

Water is ubiquitous in nature and is involved in a broad spectrum of basic and applied fields, including physics, chemistry, biology, environment, and material sciences. The interactions of water with solid surfaces have been extensively investigated, since they play a key role in a variety of scientific and technological fields, such as photocatalytic water splitting, heterogeneous and homogeneous catalysis, electrochemistry, corrosion and lubrication [1–8]. One of the most fundamental issues in all of these applied fields is the characterization of hydrogen-bonded (H-bonded) networks formed on surfaces and H-bonding dynamics in the H-boned network, which are responsible for many extraordinary physical and chemical properties of water/solid interfaces.

The complexity of water at surfaces arises from the delicate interplay between the water-water interaction and water-surface interaction, leading to a large variety of different H-bonding configurations and phases of water [3, 4, 9–14]. Water-water molecules are interconnected by H-bonding interaction. It is well known that H bonds have a strong classic component coming from electrostatics. However, its quantum component can be exceptionally prominent due to the light H nuclei (proton). The nuclear quantum effects (NQEs) of water mainly arise from quantum tunneling and zero-point motion (ZPM) of protons, which play important roles in the structure, dynamics, and macroscopic properties of H-bonded materials [15–22]. Therefore, the accurate assessment of NQEs has been a key issue for the understanding of many weirdness and the large variation in the magnitude of isotope effects of water from a quantum mechanical view.

Conventional methods of investigating interfacial water are mostly based on spectroscopic and diffraction techniques, such as sum-frequency generation (SFG), X-ray diffraction (XRD), nuclear magnetic resonance (NMR), neutron scattering and so on. However, those techniques suffer from the broadening and averaging effects due to the limitation of spatial resolution [23–25], which may easily smear out the subtle details of water-solid interaction.

A promising tool to overcome this difficulty is scanning probe microscope (SPM), including scanning tunneling microscopy (STM) and non-contact atomic force microscopy (nc-AFM), which are able to probe the solid surfaces in real space and the adsorbed molecules with Ångström resolution [26–28]. In addition to its unprecedented imaging capability, inelastic electron tunneling spectroscopy (IETS) based STM and force spectroscopy based nc-AFM enable measurement of vibrational spectroscopy and force curve, respectively, which endows SPM with chemical sensitivity allowing molecular identification [29–31] and discrimination of different species of atoms [32]. Besides using the probe tip for measurement, SPM offers the fascinating possibility of atomic manipulation techniques [33, 34], which allows the characterization of molecules at various surfaces under a well-defined environment and in a well-controlled manner. The working principle, imaging, spectroscopic capabilities and applications of STM and nc-AFM are introduced in detail in Chap. 2.

In this chapter, I first review the previous STM studies of structure and dynamics of water on metal and metal oxides surfaces. Then, I introduce the physical picture of NQEs and the impact of NQEs on water and other H-bonded systems. At last, the structure of this thesis in presented.

1.1 Water-Solid Interface

1.1.1 Structure of Water on Metal Surfaces

As a kind of model substrate, the configuration of H-bonding networks formed on the close-packed metal surfaces have been extensively investigated with STM. Individual water molecules could be visualized as a round shaped protrusion (Fig. 1.1a) [35–38] sitting on the atop site of the metal atoms of the substrate with a likely "lying" adsorption configuration (Fig. 1.1b, c), that is, the plane of the water molecule is almost parallel to the metal surface [37, 39, 40]. The imaging of water monomer is achieved at low temperature in the STM and with very low coverage due to its high mobility. When increasing the coverage, water clusters (dimer, trimer, tetramer and hexamer) are observed on the metal surfaces. Water dimer, the simplest water cluster, is identified on Pt(111) and Cu(110) with donor-acceptor configuration (Fig. 1.1d–f) [35, 41]. Density functional theory (DFT) calculations predict that the H bond donor water molecule forms stronger bond with the substrate than the acceptor, which induces the donor ~0.5 Å closer to the substrate than the acceptor (Fig. 1.1e, f) [42].

The STM image of water dimer on Pt(111) shows a "flower" like feature resulting from the model that the acceptor molecule rotates around the fixed donor molecule (Fig. 1.1d) [35]. However, the STM image of water dimer on Cu(110) appears asymmetric egg shape character with fluctuation, indicating the exchange of their roles of the molecules as donor or acceptor via H bond rearrangement during scanning [41]. Water trimers and tetramers on Cu(110) are investigated, on which the trimers prefer to form chain structure and the tetramer with cyclic structure are more favorable and stable [10, 37]. Water hexamers, which are regarded as the smallest piece of ice, are resolved with STM on many metal surfaces [43–48]. The six lobe feature

1.1 Water-Solid Interface

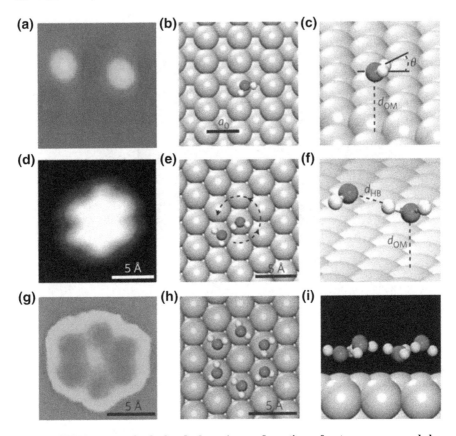

Fig. 1.1 STM images and calculated adsorption configurations of water monomer and clusters on various metal surfaces. **a–c** Water monomers on Cu(110). **d–f** Water dimer on Pt(111). **g–i** Water hexamer on Cu(111). Reprinted with permission from: **a** Ref. [41]; **d** Ref. [35]; **g** Ref. [43]; **b–i** Ref. [11]

of the hexamer STM image reveals that the water molecules form cyclic H-bonding configuration (Fig. 1.1g). DFT calculations suggest that water hexamers on reactive surfaces, such as Ru(0001), are planar because of relatively strong interaction between the water molecules and the substrate [49]. However, water hexamers are predicted to be bulked structure on noble metals (Cu and Ag) in order to balance the water-water and water-substrate interactions (Fig. 1.1h, i) [43].

At higher water coverage, water molecules aggregated and nucleated into bigger water clusters, one dimensional long chains, ice flakes and eventually two dimensional (2D) extended ice structure with increasing dosing/annealing temperature. Water molecules form hexagonal honeycomb structure on Pd(111) and Ru (0001) with additional molecules attached on the edges (periphery) of water clusters at 50 K [50]. However, the edge-molecules are metastable and disappeared above 80 K on Pd(111). When heating the Pd substrate to 130 K, the honeycomb clusters are enlarged

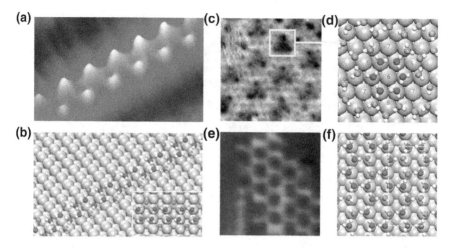

Fig. 1.2 STM images and structural model of water chains and overlayers. a, b Ice chains on Cu(110). **c, d** Water adlayer on Pt(111). **e, f** Water–hydroxyl overlayer on Cu(110). Reprinted with permission from: **a, b, d, f** Ref. [11]; **c** Ref. [54]; **e** Ref. [55]

but still with limited size [50]. On Ru(0001), below 130 K water molecules form intact stirps like chains with hexagonal structures. Once the annealing temperature is increased to 145 K, the ice structure will dissociate and substantial molecular arrangement happens, leading to the formation of mixed water-hydroxyl hexagonal strips [51]. Water molecules are self-assembled into one-dimensional chains on Cu(110) substrate with the zigzag character in the STM images (Fig. 1.2a) [52, 53]. Instead of the early suggested hexagonal structure, the long chains are proven to be built up with water pentagons in a face sharing arrangement configuration (Fig. 1.2b) [53].

With more water exposure and higher annealing temperature, 2D water layers form on some metal substrates. Combining the high resolution STM images of water layers and DFT calculations suggest that the wetting layer on Pt(111), Ru(0001) and Ni(111) are a mixture of pentagon, hexagon and heptagon water ring clusters (Fig. 1.2c, d), which totally break the bilayer ice rule [14, 54, 56]. In addition, on Cu(110) above 140 K, water molecules form distorted hexagonal H bond network, which is composed of intact water and hydroxyl groups (Fig. 1.2e, f) [55, 57]. The mixed OH and H_2O structure is further demonstrated and visualized with the nc-AFM technique [57]. Furthermore, the hexagonal and cubic ice are formed on Pt(111) surface at low temperature (140–145 K) [58]. The complex relationship between these two ice phases is elucidated by providing real-space information at the molecular scale with the STM and AFM.

1.1.2 Adsorption of Water on Metal Oxide Surfaces

The interaction of water with metal oxides has initiated broad attention and intensive studies in recent years, because metal oxides are widely used in heterogeneous catalysts, photocatalytic water splitting processes [5, 6, 8]. In comparison with the water on metals, the adsorption of water on metal oxides surfaces are more complex and less studied with microscopy methods, because the water molecules can bond not only with the metal cations but also with the oxygen anions. TiO_2 is well known an important kind of photocatalytic material, so much effort has been focused on the adsorption, diffusion and dissociation of water molecules at its surfaces. Water monomer adsorbed on the anatase $TiO_2(101)$ is revealed as "bright-dark-bright" feature at 190 K in the STM images [59]. DFT calculation predicted that the water oxygen forms a dative bond with a surface Ti_{5c} and the hydrogens form two weak H bonds with the neighbouring bridging oxygens. The $RuO_2(110)$ surface is also very interesting, because it has the similar structure with $TiO_2(110)$ and extensively used for photocatalytic water splitting. Water monomers adsorb on the Ru sites and diffuse along the Ru rows above 238 K to form dimers [60]. Increasing the temperature to 277 K, the dimer will dissociate(deprotonated) to form Ru-bound H_3O_2 and bridging hydroxyl species [60]. Isolated water molecules were observed on the Ag supported MgO films and were imaged as bright spheres residing on the top site of the magnesium cation of the surface [61].

With increasing water exposure, the formation of water clusters and ordered overlayers have been investigated by STM on the iron oxide surfaces, oxidized Cu(111) [62–65]. Combined constant current STM images and dI/dV mapping provides high resolution and comprehensive information of water clusters nucleated on the oxidized Cu(111) at room temperature and reveal that the H-bonded cyclic water clusters forms bond with the metal and oxygen cooperatively [65]. Water molecules forms large 2D islands on the bare FeO/Pt(111) surface after annealing to 130 K, while on the hydroxylated FeO surface, water molecules are inclined to form H bonds with the hydroxyl groups resulting in ordered ice-like hexameric nanoclusters at 110 K [62]. However, the wetting behavior of the α-$Fe_2O_3(0001)$ single crystal is quite different. After proper sputtering and annealing, the surface of single crystal with Fe-terminated $Fe_3O_4(111)$ and O-terminated FeO(111) domains are prepared. Water species are observed only on the Fe-terminated $Fe_3O_4(111)$ surface at 235 K, whereas there's no water molecules adsorbed on the O-terminated FeO(111) surface. What's more, when the temperature increases to 245 K, the intact water will dissociate and regular hydroxyl groups appeared on the top site of the Fe cations [64].

Recently, the interaction of water with perovskite-type oxides has received much attention as well, as perovskite-type materials are served as electrodes in a wide variety of energy conversion devices [66, 67] and have been used in electrocatalytic reactions such as water splitting or the oxygen reduction reaction [68, 69]. Isolated water molecules on the SrO-terminated surface of ruthenates dissociate, resulting in a pair of hydroxyl groups (Fig. 1.3a) [70]. At higher coverage, water monolayer formed on the Sr_2RuO_4 surface containing a mixture of dissociated OH groups and molecular

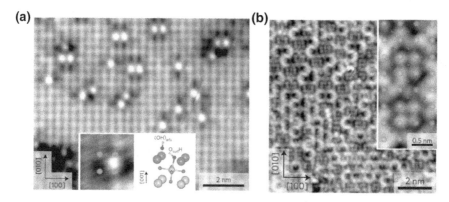

Fig. 1.3 Water adsorption at the SrO surface of $Sr_3Ru_2O_7$. a STM images of water monomer. Inset show the magnified STM image and adsorption geometry of water monomer on SrO. The water monomer dissociates and forms a pair of hydroxyl groups. **b** STM image of water monolayer. Inset is the zoom in STM image of mixed water and hydroxyl overlayer with the substrate atoms overlaid (blue—Sr, red—O). Reprinted with permission from Ref. [70]

water, with the OH adsorbed on the Sr-Sr bright site and H_2O on the top site of Sr (Fig. 1.3b) [70]. While the wetting of $Ca_3Ru_2O_7(001)$ [71] shows different behavior, on one hand, water molecules dissociate totally when annealing to room temperature, forming ordered hydroxyl network. On the other hand, the OH overlayers have a very complex phase diagram, containing (2×1), $c(2 \times 6)$, (1×3), and (1×1) periodicity because of rotation and tilt of the O octahedral in perovskites [71].

1.1.3 Dynamics of Water on Solid Surfaces

Using the STM tip, individual water molecules could be manipulated to construct water clusters. On the Cu(110) surface, one-dimensional water chain ranging from dimer to hexamer are assembled at 6 K according to the following sequence [10, 41]. First, the tip was positioned on top of the monomer; Second, decreasing the tip height so as to increase the tip-water interaction; Third, the tip was moved along the Cu rows to the vicinity of another water cluster. Finally, the tip was retracted to the initial set point and the image was rescanned. The molecule moved with the tip to the desired position and bigger water cluster formed. By manipulating water molecules and atomic oxygen, hydroxyl dimers and water-hydroxyl chains could be reproducibly assembled on Cu(110) (Fig. 1.4) [72, 73].

When electrons tunnel through the molecule, electronic and vibrational excitation will result in hopping, dissociation, desorption and other reactions [74–76]. This is especially true for water molecules. Water diffusion [47, 61, 77, 78], dissociation [36, 61, 78] desorption [78] and restructuring [77, 78] have been observed on metal and even insulating surfaces by injecting inelastic electrons from the tip. STM exper-

1.1 Water-Solid Interface

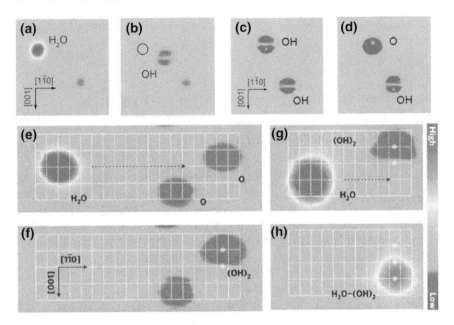

Fig. 1.4 The construction of water-hydroxyl chains on Cu(110). a–d Dissociation of a water monomer on Cu(110) by applying a voltage pulse, leading to the formation of hydroxyl **b** and oxygen atom **d**. e–h Formation sequence of water-hydroxyl chain by tip manipulation. The white grid lines indicate the lattice of Cu(110). The water molecule is manipulated along the dashed arrow. Reprinted with permission from: **a–d** Ref. [36]; **e–h** Ref. [73]

iments have proven that the diffusion of water monomer and clusters were induced and accelerated once the bending vibration mode and O–H stretching vibration mode are excited by tunneling electrons [61, 77]. When the tip is positioned over an isolated water monomer and ~2 V magnitude voltage pulse is applied, water molecules on the Ru(0001) [78] and Cu(110) [36] will dissociate producing a hydroxyl (OH) group. The hydroxyl species could be further dissociated into atomic oxygen when a voltage pulse is applied [36, 78]. In particular, Shin et al. found that the dissociation pathway of water monomers on the MgO(001) surface can be controlled by vibrational and electronic excitation, leading to different dissociated products [61]. In addition to water monomers, 3D amorphous ice could decompose and crystallize to form 2D ice clusters when irradiated by high energy tunneling electrons [79]. The large-scale dissociation of the amorphous ice suggested that the reaction was basically induced by electron injection into the conduction band of ice [79].

More recently, STM has shown the possibility of probing and manipulating the proton dynamics within the H-bonded water clusters. Kumagai et al. [73] have demonstrated that they are able to probe the proton transfer along a water-hydroxy chain that is assembled on Cu(110) by recording the tunneling current with the tip positioned over the chain [80]. Further investigation suggested that the H-atom relay reaction is assisted by vibrational excitation of the molecule induced by the inelastic tunneling electrons. Moreover, due to the light mass of protons, proton transfer can occur through quantum tunneling [36, 41].

1.2 Nuclear Quantum Effects of H-Bonded System

1.2.1 Nuclear Quantum Effects

Water is vital to human's daily life and has been extensively investigated throughout the history of science. However, the molecular-level understanding of the structure and many unusual properties of water still remains a great challenge in spite of the persistent development of new scientific instruments and theoretical methods. The mystery of water mainly arises from the intermolecular H-bonding interaction. It is well known that H bonds have a strong classic component coming from electrostatics. However, the quantum nature of H bond could not be neglected due to the quantum motion of H nuclei (proton), which is strongly correlated to the structure, dynamics, and macroscopic properties of H-bonded materials [4, 15–22]. Therefore, it is a demanding issue to describe the NQEs accurately and quantitatively, in terms of quantum tunneling and quantum fluctuation.

Water-water molecules are interconnected through H bond. Classically, the H nuclei could transfer from one side to the other through over-barrier hopping, as shown in Fig. 1.5a. Nevertheless, proton tunneling occurs (double ended blue arrow in Fig. 1.5a) when the height and width of the potential barrier is sufficiently small and has been observed in many H-bonded systems [15, 36, 81–86]. Another quantum nature of proton is ZPM or quantum fluctuation. Unlike classical particles, which are localized in the local minimum of potential well, quantum systems (i.e. proton) constantly fluctuate at zero-point energy state due to the Heisenberg uncertainty principle, namely, ZPM. The ZPM is symmetric in a harmonic potential well, where the averaged position of the proton sits at the local minimum of potential well (Fig. 1.5b). However, in real H-bonding systems, the reaction barrier exhibits anharmonic feature, leading to the expansion of O–H covalent bond compared with the case under harmonic potential, and consequently, the strengthening of H-bonding interactions (Fig. 1.5b). Moreover, once the reaction barrier of the proton transfer is lower than the ZPE, the proton will be totally delocalized and equally shared by the oxygen atoms, leading to symmetric H bond (Fig. 1.5c).

1.2.2 NQEs of Water and Ice

The NQEs usually show up in isotopic substitution experiments, where the macroscopic properties of water change significantly when replacing hydrogen (H) atoms with heavier deuterium (D) atoms (Table 1.1), which have been explicitly discussed

1.2 Nuclear Quantum Effects of H-Bonded System

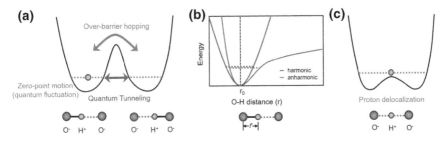

Fig. 1.5 Quantum nature of hydrogen nuclei. a Double-well potential barrier of interconversion between covalent bond and H bond. Red and blue double-ended arrows denote over-barrier hopping and quantum tunneling of proton, respectively. The zero-point motion (quantum fluctuation) of protons is depicted by dotted green lines. **b** Schematic of the harmonic (red) and anharmonic (blue) zero-point motion of the proton. r_0 denotes the equilibrium O–H distance under the harmonic potential. **c** Formation of symmetric H bond by proton delocalization. Reprinted with permission from Ref. [87]

and summarized in a recent review [22]. Both the melting temperature [88, 89] and the temperature of the maximum density [90, 91] of heavy water are higher than that of the water H_2O. Upon deuteration, the viscosity of water increases 23% accompanying with 23% decrease in water diffusion [90, 92, 93]. What's more, spectroscopic studies of liquid water suggest the picture that water D_2O is more structured than H_2O at ambient temperature [16, 94–97]. These findings provide the evidence of the weakening of H-bonding strength in liquid water at room temperature due to the impact of NQEs. However, the opposite trend appears at higher temperature [88, 98] and conflicting theoretical simulations also exist [99]. So far, whether the NQEs will enhance or decrease the H-bonding interaction is still under debate and it remains an open question what extent the quantum motion of the hydrogen nuclei can affect the hydrogen bond.

The NQEs in water/ice become more prominent at low temperature or when the separation of the nearest neighbouring oxygen atoms is small [15, 110, 111]. Conventionally, the O–O distance of water could be tuned by high pressure or structural confinement, such as confined in the nano-cavity or at interfaces. In particular, high-pressure ice has received considerable attention because of the appearance of novel behaviors. The ice experiences the phase transition from ice VIII to ice VII at high pressure, which is believed being driven by quantum tunneling of protons [15]. Since the protons in ice are usually collaborated, thus might tunnel in a correlated manner. However, the direct evidence of the concerted proton tunneling is still lacking and under debate. With further increasing pressure, the O–O distance is dramatically deceased, leading to the formation of symmetric ice X, which has been observed and confirmed by the theoretical and spectroscopic experimental approaches, where the proton is equally shared by the two nearest oxygen atoms [15, 112–114].

The adsorption of water on various solid surfaces gained extensive attention as well, as it is ubiquitous in nature and plays a crucial role in a great many environmental, biophysical, catalysis and even technological processes [1–7, 115]. The

Table 1.1 Isotope effects in water[#]

Property	H_2O	D_2O	T_2O	$H_2^{18}O$	Ref(s).
Bond energies (kJ/mol) (gas phase)	458.9	466.4 (1.6%)			[100]
Dipole moment (D) (gas phase)	1.855	1.855			[101]
Vibrational frequency (cm^{-1}) (gas phase)	3657.1	2671.6	2237.2	2237.2	[88, 102]
	1594.7	1178.4	995.4	1588.3	
	3755.9	2787.7	2366.6	3741.6	
Water dimer dissociation energy (kJ/mol) (gas phase) 10 K	13.22	14.88 (12.7%)			[103–105]
Melting point, T_m(K) (1 atm)	273.15	276.97 (1.40%)	277.64 (1.64%)	273.46 (0.11%)	[88, 89]
Temperature of maximum density (TMD) (K)	277.13	284.34 (2.60%)	286.55 (3.40%)	277.36 (0.08%)	[90, 91]
Critical temperature (K)	647.10	643.85 (−0.50%)	641.66 (−0.84%)		[88, 89]
Molar density (mol/L)	55.35	55.14 (−0.38%)	55.08 (−0.49%)	55.42 (0.13%)	[89, 106]
Molar density at the TMD (mol/L)	55.52	55.22 (−0.53%)	55.17 (−0.63%)	55.59 (0.13%)	[89, 107]
Liquid/vapor surface tension (N/m)	0.007198	0.07187 (−0.15%)			[89]
Specific heat capacity, Cv [J/(K mol)]	74.54	84.42 (13.2%)			[89, 106]
Translational diffusion ($Å^2$/ps)	0.230	0.177 (−23.0%)		0.219 (−6.1%)	[92, 93]
Rotational diffusion (rad^2/ps)	0.104	0.086 (−17%)			[108]
Dielectric relaxation time at 20 °C (ps)	9.55	12.3 (29%)			[8]
Viscosity	0.8904	1.0966 (23.2%)		0.9381 (5.4%)	[89, 90]
Acidity pH/pD	7.00	7.43	7.6		[89, 109]

Reprinted with permission from Ref. [22]
[#]Values given are at 25 °C unless otherwise stated, and values in parentheses are percentage shifts relative to H_2O

distance between adjacent water molecules is subjected to the atomic arrangement of the surfaces because of the water-substrate interaction. Using ab initio path-integral molecular dynamics (PIMD) calculation method, Li et al. unravel the substrate dependent NQEs of the H-bonding structure. The magnitude of proton delocalization in the water-hydroxyl overlayers was determined by the lattice constant of the metal substrates, leading to partially (on Pt(111) and Ru(0001)) or totally (on Ni(111)) symmetric H bonds [116]. Meanwhile, Kumagai et al. observe the symmetric H bond structure in a water-hydroxyl group on Cu(110) with the real-space STM imaging technique [72]. What's more, water dimers on Pd(111) at low temperature shows extremely high diffusion rate [47], which was supposed to be induced by H-bonding tunneling dynamics [42]. Moreover, proton tunneling processes are revealed in a water dimer on Cu(110) surface as well, resulting in the exchange of H bond donor and accepter [41].

More interestingly, water under nano-confinement shows many anomalous behaviors [117–128], such as, extremely fast proton transport through the nanochannels and exceedingly wide-range phase transition temperature of water confined in the carbon nanotube [117, 118, 120, 128]. Water confined in the carbon nanotube shows the signature of proton delocalization at low temperature [123]. More strikingly, it was recently found that a single water molecule confined inside a hexagonal shaped channel of the gemstone beryl exhibited a new quantum state, in which the proton is delocalized and tunnels between the six symmetrically equivalent positions [127].

1.2.3 NQEs in Other H-Bonded Materials and Biological Systems

Besides water, NQEs will influence the structure and properties of other H-bonded materials as well. In the recently discovered superconducting hydrogen sulfide (H_3S) system with highest superconducting transition temperature T_c of 203 K under high pressure, there is a pronounced isotope effect on T_c [129] and the superconducting phase shows the symmetric H-bonding structure with the H atoms position exactly halfway between two sulfur atoms. What's more, further investigations reveal that NQEs could influence the superconducting phase diagram of H sulfide system, in which the symmetrization pressure will be dramatically lowered when the H nuclei are treated as quantum particles, comparing with the classical case [19]. NQEs could also have a significant effect on the H-bonding interactions and structure of hydrogen fluoride (HF) systems [130, 131], protonated and hydroxylated water [110, 111].

NQEs play key roles in many biological processes such as DNA tautomerization [132–134] and enzyme catalysis reactions [135–140] as well. It was revealed that the enzyme-catalyzed reaction was dominated by proton tunneling [141] and exhibited a large kinetic isotope effect of greater than 100 [142, 143]. The enzyme ketosteroid isomerase contains a H-bonded network at its active site, which facilitates the quantum delocalization of protons, leading to a pronounced isotope effect on its acidity

[139]. Another interesting and incredible finding is that protein is more stable in D_2O compared with H_2O [144, 145] and the bacteria can survive in pure D_2O environment [146]. These observations totally renew our understanding of H-bonded systems due to the impact of NQEs. We summarize the experimental studies of NQEs with various techniques and methods in Table 1.2.

Now we can clearly see that NQEs could not just be treated as corrections to the classical H bond interactions. Instead, NQEs could have a decisive impact in the structure, dynamics, and macroscopic properties of H-bonded materials and biological systems. Although the studies of NQEs in water and aqueous systems have been summarized in several excellent reviews [17, 20–22], the accurate and quantitative description of NQEs is very scarce mainly due to the great challenge in pursuing proper treatment of the nuclear motion at a quantum mechanical level in theory and the lack of atomic-scale experimental techniques. In this thesis, I will focus on the atomic-scale investigation of NQEs of surface water with the recently advanced STM technique and PIMD calculation method.

1.3 Structure of This Thesis

In this thesis, I first introduce the previous STM studies of the structure and dynamics of water on metal and metal oxide surfaces. Then concept of NQEs and impact of NQEs on the structure, dynamics, and macroscopic properties of H-bonded systems is discussed in this chapter. As shown in Chap. 2, a detailed introduction of STM and nc-AFM is presented, focusing on the work principle, imaging, spectroscopic capabilities and applications in surface science.

In Chap. 3, using a low-temperature STM, I will report the achievement of submolecular resolution imaging of individual water monomers and tetramers adsorbed on a Au-supported NaCl(001) film at 5 K, which allows us discriminating in real space the orientation of water monomers and the H-bonding directionality of water tetramers. Based on the submolecular orbital imaging technique, we have discovered a new type of two-dimensional ice-like bilayer structure built from cyclic water tetramers on an insulating NaCl(001) film. Then I will present the nearly noninvasive submolecular-resolution imaging of water nanoclusters on a Au-supported NaCl(001) film by probing the high-order electrostatic force using a qPlus-based nc-AFM. In Chap. 4, I will focus on the measurement of vibrational spectroscopy of water. By developing tip-enhanced inelastic electron tunneling spectroscopy (IETS), we have obtained high-resolution vibrational spectroscopy of water at single molecule level, which can be employed as an ideal probe for sensing the quantum motion of protons.

Based on those novel techniques and methods, I will highlight the possibility of atomic-scale investigation of NQEs of surface water in the subsequent chapters. In Chap. 5, we have directly visualized the concerted quantum tunneling of protons within the water tetramer by monitoring the reversible interchange of H-bonding chirality in a controlled fashion with a Cl-terminated tip. The influence of the atomic-

Table 1.2 Summary of experimental studies of NQEs

Technique	System	Exhibition of NQEs	Ref(s).
Neutron scattering	Supercooled water	Proton delocalization	[147]
	Ice	Concerted proton tunneling	[86]
	Ice and water	NQEs on the proton's delocalization and vibrational properties	[148–154]
	Nanoconfined water	Proton delocalization	[123, 127, 155]
	Water in Beryl	Proton tunneling	[127]
X-ray diffraction	High pressure ice	H-bond symmetrization (ice X)	[114]
	H-bonded materials	Isotope effects on the local structure	[16, 96, 156]
	Enzyme-catalyzed reaction	Proton tunneling	[141, 157]
EELS	Hydrogen atoms on metal surfaces	Proton delocalization	[158]
Raman spectroscopy	High pressure ice	H-bond symmetrization (ice X)	[98, 113]
Infrared spectroscopy.	High pressure ice	H-bond symmetrization (ice X)	[159, 160]
	Cyclohexane on Rh(111)	NQEs on the H-bond (C-H⋯Metal) interaction	[161]
Brillouin spectroscopy	High pressure ice	H-bond symmetrization (ice X)	[162]
SFG	Interfacial water (HOD)	NQEs on the bond orientation	[163]
Rotational spectroscopy	Water hexamer prism	Concerted proton tunneling	[85]
Dielectric measurement	Ice	Concerted proton tunneling	[164]
NMR	Solid p-tert-butyl calix arene	Concerted proton tunneling	[81]
STM	Single hydrogen atoms on Cu(001)	Proton tunneling	[83]
	Heave atoms and molecules on Cu(111)	Quantum tunneling	[165–167]
	Water dimer and hydroxyl complexes on Cu(110)	Proton tunneling	[36, 41, 73]
	Water–hydroxyl complex on Cu(110)	H-bond symmetrization	[72]

(continued)

Table 1.2 (continued)

Technique	System	Exhibition of NQEs	Ref(s).
	Water tetramer on NaCl(001)	Concerted proton tunneling	[168]
	Water monomer on NaCl(001)	NQEs on the H bond strength	[169]
	Porphycene molecule on Ag(110)	Proton tunneling	[170]

Reprinted with permission from Ref. [87]

scale local environment on the concerted proton tunneling is also discussed. In Chap. 6, we have probed the quantum component of a single H bond at a water-solid interface quantitatively, revealing that the anharmonic quantum fluctuations of H nuclei weaken the weak H bonds and strengthen the relatively strong ones. However, this trend could be reversed when the H bond is strongly coupled to the atomic-scale polar environment. Finally, I will present an outlook on the current challenges and further opportunities in this field from the point of techniques and science.

References

1. Thiel PA, Madey TE (1987) The interaction of water with solid surfaces: fundamental aspects. Surf Sci Rep 7:211–385
2. Henderson MA (2002) The interaction of water with solid surfaces: fundamental aspects revisited. Surf Sci Rep 46:1–308
3. Verdaguer A, Sacha GM, Bluhm H, Salmeron M (2006) Molecular structure of water at interfaces: wetting at the nanometer scale. Chem Rev 106:1478–1510
4. Hodgson A, Haq S (2009) Water adsorption and the wetting of metal surfaces. Surf Sci Rep 64:381–451
5. Zou Z, Ye J, Sayama K, Arakawa H (2001) Direct splitting of water under visible light irradiation with an oxide semiconductor photocatalyst. Nature 414:625–627
6. Akiya N, Savage PE (2002) Roles of water for chemical reactions in high-temperature water. Chem Rev 102:2725–2750
7. Kudo A, Miseki Y (2009) Heterogeneous photocatalyst materials for water splitting. Chem Soc Rev 38:253–278
8. Eisenberg DS, Kauzmann W (1969) The structure and properties of water. Clarendon P., Oxford
9. Feibelman PJ (2010) The first wetting layer on a solid. Phys Today 63:34–39
10. Kumagai T, Okuyama H, Hatta S, Aruga T, Hamada I (2011) Water clusters on Cu(110): chain versus cyclic structures. J Chem Phys 134:024703
11. Carrasco J, Hodgson A, Michaelides A (2012) A molecular perspective of water at metal interfaces. Nat Mater 11:667–674
12. Maier S, Salmeron M (2015) How does water wet a surface? Acc Chem Res 48:2783–2790
13. Mu RT, Zhao ZJ, Dohnalek Z, Gong JL (2017) Structural motifs of water on metal oxide surfaces. Chem Soc Rev 46:1785–1806
14. Maier S, Lechner BAJ, Somorjai GA, Salmeron M (2016) Growth and structure of the first layers of ice on Ru(0001) and Pt(111). J Am Chem Soc 138:3145–3151

References

15. Benoit M, Marx D, Parrinello M (1998) Tunnelling and zero-point motion in high-pressure ice. Nature 392:258–261
16. Soper AK, Benmore CJ (2008) Quantum differences between heavy and light water. Phys Rev Lett 101:065502
17. Paesani F, Voth GA (2009) The properties of water: insights from quantum simulations. J Phys Chem B 113:5702–5719
18. Pamuk B et al (2012) Anomalous nuclear quantum effects in ice. Phys Rev Lett 108:193003
19. Errea I et al (2016) Quantum hydrogen-bond symmetrization in the superconducting hydrogen sulfide system. Nature 532:81–84
20. Marx D (2006) Proton transfer 200 years after von Grotthuss: insights from ab initio simulations. ChemPhysChem 7:1848–1870
21. Marx D, Chandra A, Tuckerman ME (2010) Aqueous basic solutions: hydroxide solvation, structural diffusion, and comparison to the hydrated proton. Chem Rev 110:2174–2216
22. Ceriotti M et al (2016) Nuclear quantum effects in water and aqueous systems: experiment, theory, and current challenges. Chem Rev 116:7529–7550
23. Shen YR, Ostroverkhov V (2006) Sum-frequency vibrational spectroscopy on water interfaces: polar orientation of water molecules at interfaces. Chem Rev 106:1140–1154
24. Nilsson A et al (2010) X-ray absorption spectroscopy and X-ray Raman scattering of water and ice; an experimental view. J Electron Spectrosc Relat Phenom 177:99–129
25. Andreani C, Colognesi D, Mayers J, Reiter GF, Senesi R (2005) Measurement of momentum distribution of light atoms and molecules in condensed matter systems using inelastic neutron scattering. Adv Phys 54:377–469
26. Repp J, Meyer G, Stojkovic SM, Gourdon A, Joachim C (2005) Molecules on insulating films: scanning-tunneling microscopy imaging of individual molecular orbitals. Phys Rev Lett 94:026803
27. Giessibl FJ (2003) Advances in atomic force microscopy. Rev Mod Phys 75:949–983
28. Gross L, Mohn F, Moll N, Liljeroth P, Meyer G (2009) The chemical structure of a molecule resolved by atomic force microscopy. Science 325:1110–1114
29. Stipe BC, Rezaei MA, Ho W (1998) Single-molecule vibrational spectroscopy and microscopy. Science 280:1732–1735
30. Stipe BC, Rezaei HA, Ho W (1999) Localization of inelastic tunneling and the determination of atomic-scale structure with chemical specificity. Phys Rev Lett 82:1724–1727
31. Chiang CL, Xu C, Han ZM, Ho W (2014) Real-space imaging of molecular structure and chemical bonding by single-molecule inelastic tunneling probe. Science 344:885–888
32. Sugimoto Y et al (2007) Chemical identification of individual surface atoms by atomic force microscopy. Nature 446:64–67
33. Eigler DM, Schweizer EK (1990) Positioning single atoms with as a scanning tunneling microscope. Nature 344:524–526
34. Stroscio JA, Eigler DM (1991) Atomic and molecular manipulation with the scanning tunneling microscope. Science 254:1319–1326
35. Motobayashi K, Matsumoto C, Kim Y, Kawai M (2008) Vibrational study of water dimers on Pt(111) using a scanning tunneling microscope. Surf Sci 602:3136–3139
36. Kumagai T et al (2009) Tunneling dynamics of a hydroxyl group adsorbed on Cu(110). Phys Rev B 79:035423
37. Okuyama H, Hamada I (2011) Hydrogen-bond imaging and engineering with a scanning tunnelling microscope. J Phys D Appl Phys 44:464004
38. Shimizu TK et al (2008) Surface species formed by the adsorption and dissociation of water molecules on a Ru(0001) surface containing a small coverage of carbon atoms studied by scanning tunneling microscopy. J Phys Chem C 112:7445–7454
39. Michaelides A, Ranea VA, de Andres PL, King DA (2003) General model for water monomer adsorption on close-packed transition and noble metal surfaces. Phys Rev Lett 90:216102
40. Meng S, Wang EG, Gao SW (2004) Water adsorption on metal surfaces: a general picture from density functional theory studies. Phys Rev B 69:195404

41. Kumagai T et al (2008) Direct observation of hydrogen-bond exchange within a single water dimer. Phys Rev Lett 100:166101
42. Ranea VA et al (2004) Water dimer diffusion on Pd(111) assisted by an H-bond donor-acceptor tunneling exchange. Phys Rev Lett 92:136104
43. Michaelides A, Morgenstern K (2007) Ice nanoclusters at hydrophobic metal surfaces. Nat Mater 6:597–601
44. Morgenstern K (2002) Scanning tunnelling microscopy investigation of water in submonolayer coverage on Ag(111). Surf Sci 504:293–300
45. Gawronski H, Carrasco J, Michaelides A, Morgenstern K (2008) Manipulation and control of hydrogen bond dynamics in absorbed ice nanoclusters. Phys Rev Lett 101:136102
46. Mehlhorn M, Carrasco J, Michaelides A, Morgenstern K (2009) Local investigation of femtosecond laser induced dynamics of water nanoclusters on Cu(111). Phys Rev Lett 103:026101
47. Mitsui T, Rose MK, Fomin E, Ogletree DF, Salmeron M (2002) Water diffusion and clustering on Pd(111). Science 297:1850–1852
48. Chen JW, Tu XY, Tian K, Dai SS (2006) Density functional theory study of water diffusion and clustering on Pd(111). Chin J Struct Chem 25:909–914
49. Haq S, Clay C, Darling GR, Zimbitas G, Hodgson A (2006) Growth of intact water ice on Ru(0001) between 140 and 160 K: Experiment and density-functional theory calculations. Phys Rev B 73:115414
50. Tatarkhanov M et al (2009) Metal- and hydrogen-bonding competition during water adsorption on Pd(111) and Ru(0001). J Am Chem Soc 131:18425–18434
51. Maier S, Stass I, Cerda JI, Salmeron M (2014) Unveiling the Mechanism of Water Partial Dissociation on Ru(0001). Phys Rev Lett 112:126101
52. Yamada T, Tamamori S, Okuyama H, Aruga T (2006) Anisotropic water chain growth on Cu(110) observed with scanning tunneling microscopy. Phys Rev Lett 96:036105
53. Carrasco J et al (2009) A one-dimensional ice structure built from pentagons. Nat Mater 8:427–431
54. Nie S, Feibelman PJ, Bartelt NC, Thuermer K (2010) Pentagons and heptagons in the first water layer on Pt(111). Phys Rev Lett 105:026102
55. Forster M, Raval R, Hodgson A, Carrasco J, Michaelides A (2011) c(2 x 2) Water-hydroxyl layer on Cu(110): a wetting layer stabilized by Bjerrum defects. Phys Rev Lett 106:046103
56. Thurmer K, Nie S, Feibelman PJ, Bartelt NC (2014) Clusters, molecular layers, and 3D crystals of water on Ni(111). J Chem Phys 141:18C520
57. Shiotari A, Sugimoto Y (2017) Ultrahigh-resolution imaging of water networks by atomic force microscopy. Nat Commun 8:14313
58. Thurmer K, Nie S (2013) Formation of hexagonal and cubic ice during low-temperature growth. Proc Natl Acad Sci USA 110:11757–11762
59. He Y, Tilocca A, Dulub O, Selloni A, Diebold U (2009) Local ordering and electronic signatures of submonolayer water on anatase $TiO_2(101)$. Nat Mater 8:585–589
60. Mu R et al (2014) Dimerization induced deprotonation of water on $RuO_2(110)$. J Phys Chem Lett 5:3445–3450
61. Shin H-J et al (2010) State-selective dissociation of a single water molecule on an ultrathin MgO film. Nat Mater 9:442–447
62. Merte LR et al (2014) Water clustering on nanostructured iron oxide films. Nat Commun 5:4193
63. Merte LR et al (2012) Water-mediated proton hopping on an iron oxide surface. Science 336:889–893
64. Rim KT et al (2012) Scanning tunneling microscopy and theoretical study of water adsorption on Fe_3O_4: implications for catalysis. J Am Chem Soc 134:18979–18985
65. Kronawitter CX et al (2014) Hydrogen-bonded cyclic water clusters nucleated on an oxide surface. J Am Chem Soc 136:13283–13288
66. Kilner JA, Burriel M (2014) Materials for intermediate-temperature solid-oxide fuel cells. Annu Rev Mater Res 44:365–393

References

67. Liu M, Winnick J (1997) Electrode kinetics of porous mixed-conducting oxygen electrodes. J Electrochem Soc 144:1881–1884
68. Vojvodic A, Norskov JK (2011) Optimizing perovskites for the water-splitting reaction. Science 334:1355–1356
69. Suntivich J, May KJ, Gasteiger HA, Goodenough JB, Shao-Horn Y (2011) A perovskite oxide optimized for oxygen evolution catalysis from molecular orbital principles. Science 334:1383–1385
70. Halwidl D et al (2016) Adsorption of water at the SrO surface of ruthenates. Nat Mater 15:450–455
71. Halwidl D et al (2017) Ordered hydroxyls on $Ca_3Ru_2O_7(001)$. Nat Commun 8:23
72. Kumagai T et al (2010) Symmetric hydrogen bond in a water-hydroxyl complex on Cu(110). Phys Rev B 81:045402
73. Kumagai T et al (2012) H-atom relay reactions in real space. Nat Mater 11:167–172
74. Komeda T, Kim Y, Kawai M, Persson BNJ, Ueba H (2002) Lateral hopping of molecules induced by excitation of internal vibration mode. Science 295:2055–2058
75. Pascual JI, Lorente N, Song Z, Conrad H, Rust HP (2003) Selectivity in vibrationally mediated single-molecule chemistry. Nature 423:525–528
76. Kim Y, Komeda T, Kawai M (2002) Single-molecule reaction and characterization by vibrational excitation. Phys Rev Lett 89:126104
77. Morgenstern K, Rieder KH (2002) Formation of the cyclic ice hexamer via excitation of vibrational molecular modes by the scanning tunneling microscope. J Chem Phys 116:5746–5752
78. Mugarza A, Shimizu TK, Ogletree DF, Salmeron M (2009) Chemical reactions of water molecules on Ru(0001) induced by selective excitation of vibrational modes. Surf Sci 603:2030–2036
79. Morgenstern K, Rieder KH (2002) Dissociation of water molecules with the scanning tunnelling microscope. Chem Phys Lett 358:250–256
80. Mehlhorn M, Gawronski H, Morgenstern K (2008) Electron damage to supported ice investigated by scanning tunneling microscopy and spectroscopy. Phys Rev Lett 101:196101
81. Brougham DF, Caciuffo R, Horsewill AJ (1999) Coordinated proton tunnelling in a cyclic network of four hydrogen bonds in the solid state. Nature 397:241–243
82. Tomchuk PM, Krasnoholovets VV (1997) Macroscopic quantum tunneling of polarization in the hydrogen-bonded chain. J Mol Struct 416:161–165
83. Lauhon LJ, Ho W (2000) Direct observation of the quantum tunneling of single hydrogen atoms with a scanning tunneling microscope. Phys Rev Lett 85:4566–4569
84. Kumagai T (2015) Direct observation and control of hydrogen-bond dynamics using low-temperature scanning tunneling microscopy. Prog Surf Sci 90:239–291
85. Richardson JO et al (2016) Concerted hydrogen-bond breaking by quantum tunneling in the water hexamer prism. Science 351:1310–1313
86. Bove LE, Klotz S, Paciaroni A, Sacchetti F (2009) Anomalous proton dynamics in ice at low temperatures. Phys Rev Lett 103:165901
87. Guo J, Li XZ, Peng JB, Wang EG, Jiang Y (2017) Atomic-scale investigation of nuclear quantum effects of surface water: experiments and theory. Prog Surf Sci 92:203–239
88. Anonymous (NIST Chemistry WebBook); NIST Standard Reference Database 69; National Institute of Standards and Technology (NIST): Gaithersburg, MD, 2005. Available at http://Webbook.Nist.gov. Accessed 19 Dec 2014
89. IAPWS Releases. Supplementary releases, guidelines, and advisory notes. International Association for the properties of water and steam (IAPWS). Available at http://www.iapws.org/release.html. Accessed 21 Dec 2014
90. Kudish AI, Steckel F, Wolf D (1972) Physical properties of heavy-oxygen water. Absolute viscosity of $H_2{}^{18}O$ between 15 and 35 °C. J Chem Soc Furuduy Trans 1 Phys Chem Condens Phases 68:2041–2046
91. Hill PG, Macmillan RDC, Lee V (1982) A fundamental equation of state for heavy-water. J Phys Chem Ref Data 11:1–14

92. Holz M, Heil SR, Sacco A (2000) Temperature-dependent self-diffusion coefficients of water and six selected molecular liquids for calibration in accurate ^{1}H NMR PFG measurements. Phys Chem Chem Phys 2:4740–4742
93. Price WS, Ide H, Arata Y, Soderman O (2000) Temperature dependence of the self-diffusion of supercooled heavy water to 244 K. J Phys Chem B 104:5874–5876
94. Root JH, Egelstaff PA, Hime A (1986) Quantum effects in the structure of water measured by gamma ray diffraction. Chem Phys 109:437–453
95. Tomberli B, Benmore CJ, Egelstaff PA, Neuefeind J, Honkimaki V (2000) Isotopic quantum effects in water structure measured with high energy photon diffraction. J Phys Condens Matter 12:2597–2612
96. Bergmann U et al (2007) Isotope effects in liquid water probed by X-ray Raman spectroscopy. Phys Rev B 76:024202
97. Harada Y et al (2013) Selective probing of the OH or OD stretch vibration in liquid water using resonant inelastic soft-X-ray scattering. Phys Rev Lett 111:193001
98. Goncharov AF, Struzhkin VV, Mao HK, Hemley RJ (1999) Raman spectroscopy of dense H_2O and the transition to symmetric hydrogen bonds. Phys Rev Lett 83:1998–2001
99. Chen B, Ivanov I, Klein ML, Parrinello M (2003) Hydrogen bonding in water. Phys Rev Lett 91:215503
100. Maksyutenko P, Rizzo TR, Boyarkin OV (2006) A direct measurement of the dissociation energy of water. J Chem Phys 125:181101
101. Dyke TR, Muenter JS (1973) Electric dipole-moments of low J states of H_2O and D_2O. J Chem Phys 59:3125–3127
102. Tennyson J et al (2014) IUPAC critical evaluation of the rotational-vibrational spectra of water vapor. Part IV. Energy levels and transition wavenumbers for $D_2^{16}O$, $D_2^{17}O$, and $D_2^{18}O$. J Quant Spectro Radiat Transfer 142:93–108
103. Rocher-Casterline BE, Ch'ng LC, Mollner AK, Reisler H (2011) Communication: determination of the bond dissociation energy (D_0) of the water dimer, $(H_2O)_2$, by velocity map imaging. J Chem Phys 134:211101
104. Rocher-Casterline BE, Mollner AK, Ch'ng LC, Reisler H (2011) Imaging H_2O photofragments in the predissociation of the $HCl-H_2O$ hydrogen-bonded dimer. J Phys Chem A 115:6903–6909
105. Ch'ng LC, Samanta AK, Czako G, Bowman JM, Reisler H (2012) Experimental and theoretical investigations of energy transfer and hydrogen-bond breaking in the water dimer. J Am Chem Soc 134:15430–15435
106. Nakamura M, Tamura K, Murakami S (1995) Isotope effects on thermodynamic properties-mixtures of $x(D_2O$ or $H_2O)+(1-x)CH_3CN$ at 298.15 K. Thermochim Acta 253:127–136
107. Kell GS (1977) Effects of isotopic composition, temperature, pressure, and dissolved-gases on density of liquid water. J Phys Chem Ref Data 6:1109–1131
108. Hardy EH, Zygar A, Zeidler MD, Holz M, Sacher FD (2001) Isotope effect on the translational and rotational motion in liquid water and ammonia. J Chem Phys 114:3174–3181
109. Lide DR (ed) (1999) CRC handbook of chemistry and physics, 80th edn. CRC Press, Boca Raton
110. Tuckerman ME, Marx D, Klein ML, Parrinello M (1997) On the quantum nature of the shared proton in hydrogen bonds. Science 275:817–820
111. Marx D, Tuckerman ME, Hutter J, Parrinello M (1999) The nature of the hydrated excess proton in water. Nature 397:601–604
112. Hirsch KR, Holzapfel WB (1984) Symmetric hydrogen bonds in ice X. Phys Lett A 101:142–144
113. Hirsch K, Holzapfel W (1986) Effect of high pressure on the Raman spectra of ice VIII and evidence for ice X. J Chem Phys 84:2771–2775
114. Loubeyre P, LeToullec R, Wolanin E, Hanfland M, Husermann D (1999) Modulated phases and proton centring in ice observed by X-ray diffraction up to 170 GPa. Nature 397:503–506
115. Israelachvili J, Wennerstrom H (1996) Role of hydration and water structure in biological and colloidal interactions. Nature 379:219–225

References

116. Li X-Z, Probert MIJ, Alavi A, Michaelides A (2010) Quantum nature of the proton in water-hydroxyl overlayers on metal surfaces. Phys Rev Lett 104:066102
117. Roux B, Karplus M (1991) Ion-Transport in a Gramicidin-like channel—dynamics and mobility. J Phys Chem 95:4856–4868
118. Pomes R, Roux B (1996) Structure and dynamics of a proton wire: a theoretical study of H^+ translocation along the single-file water chain in the gramicidin A channel. Biophys J 71:19–39
119. Koga K, Gao GT, Tanaka H, Zeng XC (2001) Formation of ordered ice nanotubes inside carbon nanotubes. Nature 412:802–805
120. Dellago C, Naor MM, Hummer G (2003) Proton transport through water-filled carbon nanotubes. Phys Rev Lett 90:105902
121. Kolesnikov AI et al (2004) Anomalously soft dynamics of water in a nanotube: a revelation of nanoscale confinement. Phys Rev Lett 93:035503
122. Maniwa Y et al (2005) Ordered water inside carbon nanotubes: formation of pentagonal to octagonal ice-nanotubes. Chem Phys Lett 401:534–538
123. Reiter G et al (2006) Anomalous behavior of proton zero point motion in water confined in carbon nanotubes. Phys Rev Lett 97:247801
124. Garbuio V et al (2007) Proton quantum coherence observed in water confined in silica nanopores. J Chem Phys 127:154501
125. Reiter GF et al (2013) Anomalous ground state of the electrons in nanoconfined water. Phys Rev Lett 111:036803
126. Algara-Siller G et al (2015) Square ice in graphene nanocapillaries. Nature 519:443–445
127. Kolesnikov AI et al (2016) Quantum tunneling of water in beryl: a new state of the water molecule. Phys Rev Lett 116:167802
128. Agrawal KV, Shimizu S, Drahushuk LW, Kilcoyne D, Strano MS (2017) Observation of extreme phase transition temperatures of water confined inside isolated carbon nanotubes. Nat Nanotech 12:267–273
129. Drozdov AP, Eremets MI, Troyan IA, Ksenofontov V, Shylin SI (2015) Conventional superconductivity at 203 kelvin at high pressures in the sulfur hydride system. Nature 525:73–76
130. Raugei S, Klein ML (2003) Nuclear quantum effects and hydrogen bonding in liquids. J Am Chem Soc 125:8992–8993
131. Swalina C, Wang Q, Chakraborty A, Hammes-Schiffer S (2007) Analysis of nuclear quantum effects on hydrogen bonding. J Phys Chem A 111:2206–2212
132. Douhal A, Kim SK, Zewail AH (1995) Femtosecond molecular dynamics of tautomerization in model base pairs. Nature 378:260–263
133. Kwon OH, Zewail AH (2007) Double proton transfer dynamics of model DNA base pairs in the condensed phase. Proc Natl Acad Sci USA 104:8703–8708
134. Perez A, Tuckerman ME, Hjalmarson HP, von Lilienfeld OA (2010) Enol tautomers of watson-crick base pair models are metastable because of nuclear quantum effects. J Am Chem Soc 132:11510–11515
135. Hwang JK, Warshel A (1996) How important are quantum mechanical nuclear motions in enzyme catalysis? J Am Chem Soc 118:11745–11751
136. Billeter SR, Webb SP, Agarwal PK, Iordanov T, Hammes-Schiffer S (2001) Hydride transfer in liver alcohol dehydrogenase: quantum dynamics, kinetic isotope effects, and role of enzyme motion. J Am Chem Soc 123:11262–11272
137. Pu JZ, Gao JL, Truhlar DG (2006) Multidimensional tunneling, recrossing, and the transmission coefficient for enzymatic reactions. Chem Rev 106:3140–3169
138. Glowacki DR, Harvey JN, Mulholland AJ (2012) Taking Ockham's razor to enzyme dynamics and catalysis. Nat Chem 4:169–176
139. Wang L, Fried SD, Boxer SG, Markland TE (2014) Quantum delocalization of protons in the hydrogen-bond network of an enzyme active site. Proc Natl Acad Sci USA 111:18454–18459
140. Truhlar DG (2010) Tunneling in enzymatic and nonenzymatic hydrogen transfer reactions. J Phys Org Chem 23:660–676

141. Masgrau L et al (2006) Atomic description of an enzyme reaction dominated by proton tunneling. Science 312:237–241
142. Sutcliffe MJ, Scrutton NS (2002) A new conceptual framework for enzyme catalysis—hydrogen tunneling coupled to enzyme dynamics in flavoprotein and quinoprotein enzymes. Eur J Biochem 269:3096–3102
143. Klinman JP, Kohen A (2013) Hydrogen tunneling Links protein dynamics to enzyme catalysis. Annu Rev Biochem 82(82):471–496
144. Efimova YM, Haemers S, Wierczinski B, Norde W, van Well AA (2007) Stability of globular proteins in H_2O and D_2O. Biopolymers 85:264–273
145. Cho Y et al (2009) Hydrogen bonding of beta-Turn structure is stabilized in D_2O. J Am Chem Soc 131:15188–15193
146. Mosin OV, Shvets VI, Skladnev DA, Ignatov I (2014) Studying of microbic synthesis of deuterium labelled L-Phenylalanine by facultative methylotrophic bacterium brevibacterium methylicum on media with different content of heavy water. Biomeditsinskaya Khimiya 60:448–461
147. Pietropaolo A et al (2008) Excess of proton mean kinetic energy in supercooled water. Phys Rev Lett 100:127802
148. Senesi R et al (2013) The quantum nature of the OH stretching mode in ice and water probed by neutron scattering experiments. J Chem Phys 139:074504
149. Senesi R, Romanelli G, Adams MA, Andreani C (2013) Temperature dependence of the zero point kinetic energy in ice and water above room temperature. Chem Phys 427:111–116
150. Zeidler A et al (2011) Oxygen as a site specific probe of the structure of water and oxide materials. Phys Rev Lett 107:145501
151. Romanelli G et al (2013) Direct measurement of competing quantum effects on the kinetic energy of heavy water upon melting. J Phys Chem Lett 4:3251–3256
152. Burnham CJ et al (2006) On the origin of the redshift of the OH stretch in Ice Ih: evidence from the momentum distribution of the protons and the infrared spectral density. Phys Chem Chem Phys 8:3966–3977
153. Burnham CJ, Anick DJ, Mankoo PK, Reiter GF (2008) The vibrational proton potential in bulk liquid water and ice. J Chem Phys 128:154519
154. Pantalei C et al (2008) Proton momentum distribution of liquid water from room temperature to the supercritical phase. Phys Rev Lett 100:177801
155. Reiter GF et al (2012) Evidence for an anomalous quantum state of protons in nanoconfined water. Phys Rev B 85:045403
156. Ubbelohde AR, Gallagher KJ (1955) Acid-base effects in hydrogen bonds in crystals. Acta Crystallogr 8:71–83
157. Major DT et al (2009) Differential quantum tunneling contributions in nitroalkane oxidase catalyzed and the uncatalyzed proton transfer reaction. Proc Natl Acad Sci USA 106:20734–20739
158. Nishijima M, Okuyama H, Takagi N, Aruga T, Brenig W (2005) Quantum delocalization of hydrogen on metal surfaces. Surf Sci Rep 57:113–156
159. Aoki K, Yamawaki H, Sakashita M, Fujihisa H (1996) Infrared absorption study of the hydrogen-bond symmetrization in ice to 110 GPa. Phys Rev B 54:15673–15677
160. Goncharov AF, Struzhkin VV, Somayazulu MS, Hemley RJ, Mao HK (1996) Compression of ice to 210 gigapascals: infrared evidence for a symmetric hydrogen-bonded phase. Science 273:218–220
161. Koitaya T, Yoshinobu J (2014) The quantum nature of C-H···Metal interaction: vibrational spectra and kinetic and geometric isotope effects of adsorbed cyclohexane. Chem Rec 14:848–856
162. Polian A, Grimsditch M (1984) New high-pressure phase of H_2O: ice X. Phys Rev Lett 52:1312–1314
163. Nagata Y, Pool RE, Backus EHG, Bonn M (2012) Nuclear quantum effects affect bond orientation of water at the water-vapor interface. Phys Rev Lett 109:226101

164. Yen F, Gao T (2015) Dielectric anomaly in ice near 20 K: evidence of macroscopic quantum phenomena. J Phys Chem Lett 6:2822–2825
165. Heinrich AJ, Lutz CP, Gupta JA, Eigler DM (2002) Molecule cascades. Science 298:1381–1387
166. Repp J, Meyer G, Rieder KH, Hyldgaard P (2003) Site determination and thermally assisted tunneling in homogenous nucleation. Phys Rev Lett 91:206102
167. Stroscio JA, Celotta RJ (2004) Controlling the dynamics of a single atom in lateral atom manipulation. Science 306:242–247
168. Meng X et al (2015) Direct visualization of concerted proton tunnelling in a water nanocluster. Nat Phys 11:235–239
169. Guo J et al (2016) Nuclear quantum effects of hydrogen bonds probed by tip-enhanced inelastic electron tunneling. Science 352:321–325
170. Koch M et al (2017) Direct observation of double hydrogen transfer via quantum tunneling in a single porphycene molecule on a Ag(110) surface. J Am Chem Soc 139:12681–12687

Chapter 2
Scanning Probe Microscopy

Based on the concept of quantum tunneling, scanning tunneling microscopy (STM) [1–3] was invented by Gerd Binnig and Heinrich Rohrer at IBM Zürich in 1981. Using STM, they achieved the atomic structure of metal and the famous Si(111) surface with 7×7 reconstruction [4]. Since "seeing is believing", STM is proven to be one of the most versatile and important tools in the surface science research field, which owns the ability of real space imaging with sub-Ångström resolution [5]. Despite the phenomenal success of the STM, it usually requires conductive substrates or atomically thin insulating layers, which greatly limits the application of STM.

To overcome this limitation, atomic force microscopy (AFM) [6] was invented by Gerd Binnig et al. in 1986, which operates by measuring the interaction force between the sharp AFM tip and the sample surface. As the electrical conductivity of the sample is not required in AFM, the AFM has been developed to work not only in the ultrahigh vacuum (UHV), ambient environment but also in the liquid. As a result, the AFM has been applied to investigate problems in a wide range of disciplines of the natural sciences, including solid-state physics, semiconductor science and technology, molecular engineering, polymer chemistry and physics, surface chemistry, molecular biology, cell biology, and medicine.

Besides AFM, a series of related techniques have been developed, such as ballistic electron emission microscopy (BEEM) [7], kelvin probe force microscopy (KPFM) [8], magnetic force microscopy (MFM) [9], scanning near-field optical microscopy (SNOM) [10], scanning microwave impedance microscopy (sMIM) [11] and so on. All these techniques are summarized as scanning probe microscopy (SPM) because they have in common that using a scanning probe to image and measure locally physical and chemical properties down to the level of atomic-scale.

In this chapter, I will give a detailed introduction of STM and non-contact atomic force microscopy (nc-AFM), focusing on the work principle, imaging, spectroscopic capabilities and applications in surface science.

2.1 Scanning Tunneling Microscopy

2.1.1 Principle

STM is based on the several principles. One is the quantum mechanical effect of tunneling. As shown in Fig. 2.1, STM contains a sharp metal tip and a conducting sample. A voltage bias is applied between the tip and the sample. When the conducting tip is approached very near to the surface of the substrate (<1 nm), the electrons could tunnel through the vacuum barrier and creates tunneling current. It is this quantum effect allows us to "see" the surface. Another principle is the piezoelectric effect, which allow us to precisely control the tip in xyz directions at the Ångström-level. Lastly, a feedback loop is required, which monitors the tunneling current and makes adjustments to the tip to maintain a constant tunneling current. The tunneling gap between the STM probe and the sample can be tuned by adjusting the bias and tunneling current when the feedback loop is closed.

STM has two working modes [3], constant height mode and constant current mode. In the constant height mode, the tip height and bias voltage are both constant while the current changes when the tip scanning the surface, which is related to the topography and charge density of the surface. In the constant current mode, the feedback electronics is switched on and the tip is adjusted to maintain the current constant. The contrast in the STM image (variation of the tip height) show the topography of the surface directly, accompanied with the information of surface density

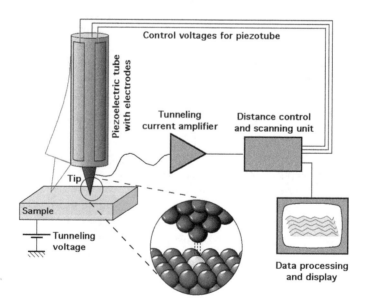

Fig. 2.1 Schematic view of STM

2.1 Scanning Tunneling Microscopy

state. Comparing with the constant height mode, constant current mode is much safer and is widely used in the experiment.

Then, we discuss the concept of tunneling. In classical mechanics, an object cannot pass through a potential barrier if its energy E is smaller than potential of the barrier U. In contrast, in quantum mechanics, objects with a very small mass, such as the electron, which is treated as a wavefunction $\psi(z)$, has a non-zero possibility of penetrating through the barrier. For a rectangular barrier, a quantum mechanical treatment predicts an exponential decaying solution for the electron wave function in the barrier

$$\psi(z) = \psi(0)e^{-\kappa z} \tag{2.1}$$

where $\kappa = \frac{\sqrt{2m(E-U)}}{\hbar}$.

The probability of observing an electron behind the barrier of the width z is

$$\left|\psi(z)^2\right| = \left|\psi(0)^2\right|e^{-2\kappa z} \tag{2.2}$$

Starting from this elementary model, we can explain some basic feature of STM. The workfunction Φ of a metal surface is defined as the minimum energy required to remove an electron form the bulk to the vacuum level. In STM, a small voltage pulse V is applied between the tip and sample and results in a tunneling current I. The height of the barrier can roughly be approximated by the average workfunction of sample and tip.

$$\Phi = \frac{1}{2}\left(\Phi_{\text{sam}} + \Phi_{tip}\right) \tag{2.3}$$

If the bias is much smaller the value of the workfunction, that is, $eV \ll \Phi$. The inverse decay length κ can be simplified to

$$\kappa = \frac{\sqrt{2m\Phi}}{\hbar} \tag{2.4}$$

The tunneling current is proportional to the probability of electrons to tunnel through the barrier:

$$I \propto \left|\psi(0)^2\right|e^{-2\kappa z} \tag{2.5}$$

The typical value of workfunction of metal materials is about $\Phi \approx 5\,\text{eV}$ [12], which gives the value of the decay constant $\kappa \approx 11.4\,\text{nm}^{-1}$. According to the Eq. (2.5), the tunneling current decays one order of magnitude when the tip sample distance varies 1 Å, which results in the high vertical resolution of STM.

Another way of describing electron tunneling is the time dependent perturbation theory developed by Bardeen [13]. Instead of solving the Schrödinger's equation of combined system, Bardeen solved Schrödinger's equation for each side of the metal-insulator-metal junction separately to obtain the wave functions for each electrode. This can be applied to STM, make the electrodes as tip and sample. The wavefunctions of tip and sample are assigned as ψ_t and ψ_s respectively. The tunneling matrix element M_{st} denotes the overlap of the tip and sample wave functions. The tunneling current is

$$I = \frac{2\pi e^2}{\hbar}|M_{st}|^2 \rho_s(E_F)\rho_t(E_F) \tag{2.6}$$

where

$$M_{st} = \frac{\hbar^2}{2m}\int d\mathbf{S} \cdot \left(\psi_t^*\nabla\psi_s - \psi_s^*\nabla\psi_t\right) \tag{2.7}$$

$\rho_s(E)$ and $\rho_t(E)$ are the density of states (DOS) of sample and tip at energy E, respectively. With a bias voltage V_T, the total tunneling current [14] is

$$I_T = \frac{4\pi e}{\hbar}\int_{-\infty}^{\infty}[f_t(E_F - eV_T + \varepsilon) - f_s(E_F + \varepsilon)]$$
$$\times \rho_t(E_F + \varepsilon - eV_T)\rho_s(E_F + \varepsilon)|M(E_F + \varepsilon - eV_T, \varepsilon)|^2 d\varepsilon \tag{2.8}$$

where the Fermi-Dirac distribution $f(E) = (1 + \exp[(E - E_F)/k_B t])^{-1}$. As it can be seen from the Eq. (2.8), the tunneling current depends not only on the DOS of tip and sample, but also on the tunneling matrix element M_{st}. However, it is difficult to determine the tunneling matrix element in the realistic case, which depends on the geometric structure of the tip and sample as well as the wave functions of the two electrodes. Unfortunately, in most cases the geometric and chemical structure of the tip is unknow. Using first-order perturbation theory, Tersoff and Hamann [15, 16] presented an analytical result for the tunneling matrix element, which is suitable for the atomically sharp tip, where only the last few atoms close to the sample surface at the tip apex contribute to the tunneling process. In such case, the tip apex is described as spherical, s-like orbital and the DOS of the tip is considered to be constant. In the low voltage bias region $|V_T| \ll \Phi/e$, the tunneling matrix element is written as,

$$M_{\mu\nu} = -\frac{2\pi C\hbar^2}{\kappa m_e}\psi_s(\vec{r}) \tag{2.9}$$

where C is a constant, $\vec{r}$ is the position of the center of curvature of the tip. In the Tersoff-Hamann model, the tunneling matrix only depends on the wavefunction of sample at the position of $\vec{r}$, not on the wavefunction of the tip. Then the total tunneling current at T $\to$ 0 is

2.1 Scanning Tunneling Microscopy

$$I_T = \frac{16\pi^3 C^2 \hbar^3 e}{\kappa^2 m_e^2} \rho_t \int_0^{eV_T} \rho_s(\varepsilon) d\varepsilon, \tag{2.10}$$

which is proportional to the DOS of the sample.

2.1.2 Scanning Tunneling Spectroscopy

Besides high resolution imaging, STM also owns the capability of measuring the spectroscopic data of the sample surface at the atomic scale. In this section, I will introduce scanning tunneling spectroscopy (STS), which is used to provide information about the density of electrons in a sample as a function of their energy. Generally, STS involves local measurement of the derivative of the tunneling current versus tip-sample bias, that is, the tunneling conductance, dI/dV. With the approximation that tunneling matrix is constant, tunneling current expression 2.8 can be written as

$$I_T \propto \int_{-\infty}^{\infty} \rho_t(\varepsilon - eV_T)\rho_s(\varepsilon)(f_t(\varepsilon - eV_T) - f_s(\varepsilon))d\varepsilon \tag{2.11}$$

We calculate the first derivative of the tunneling current I_T with respect to the applied bias voltage V:

$$\frac{\partial I_T}{\partial V}|_{V_T} \propto \int_{-\infty}^{\infty} d\varepsilon \big[-\rho_s(\varepsilon)\rho_t'(\varepsilon - eV_T)f_t(\varepsilon - eV_T) - \rho_s(\varepsilon)\rho_t(\varepsilon - eV_T)f_t'(\varepsilon - eV_T)$$

$$+\rho_s(\varepsilon)\rho_t'(\varepsilon - eV_T)f_t(\varepsilon - eV_T)f_s(\varepsilon)\big] \tag{2.12}$$

where $\rho_t'(\varepsilon) = \frac{\partial \rho_t(\varepsilon)}{\partial \varepsilon}$

$$f'(\varepsilon) = \frac{\partial f(\varepsilon)}{\partial \varepsilon} = \frac{-\exp(\varepsilon/k_B T)}{k_B T(1 + \exp(\varepsilon/k_B T))^2} = -\frac{1}{2k_B T}\text{sech}^2(\varepsilon/k_B T) \tag{2.13}$$

To simplify the Eq. (2.12), we assume that the LDOS of the tip is constant ($\rho_t' = 0$) and the temperature of the STM junction is zero (T → 0). In such case, the first derivative of the Fermi-Dirac distribution function becomes the delta distribution, $f'(\varepsilon)_{k_B T \to 0} = -\delta(\varepsilon)$. Then the first derivative of the tunneling current can be written as

$$\frac{\partial I_T}{\partial V}|_{V_T} \propto \rho_t \int_{-\infty}^{\infty} \rho_s(\varepsilon)\delta(\varepsilon - eV_T)d\varepsilon = \rho_t \rho_s(eV_T) \tag{2.14}$$

As a result, the tunneling conductance is directly proportional to the sample LDOS with these ideal assumptions. In the experiment, STS is obtained by positioning a STM tip above a particular place on the sample. With the height of the tip fixed, the electron tunneling current is then measured as a function of electron energy by ramping the bias voltage between the tip and the sample. The STS data can be calculated numerically from the $I(V_T)$ curve. However, with the numerical calculation method, the STS curve is usually very noise. Then Lock-in technique is used to detect the STS data directly with high signal to noise ratio.

In lock-in technique, the DC bias voltage V_T is modulated with a small, high frequency sinusoidal voltage $V_m \sin(\omega_m t)$. The modulation frequency is thereby set to much higher values as the regulation speed of the feedback loop that holds the tunneling current constant in closed-loop mode, thus to make sure that the bias modulation does not influence the recording of constant-current images. The AC component of the tunneling current is recorded using a lock-in amplifier, and the component in-phase with the tip-sample bias modulation gives dI/dV directly.

Based on the Tersoff-Hamann model, the tunneling current can be described as

$$I_T \propto \int_0^{eV_T + V_m \sin(\omega_m t)} \rho_s(\varepsilon) d\varepsilon \tag{2.15}$$

Expanding the current in a Taylor series:

$$I_T \propto \int_0^{eV_T} \rho_s(\varepsilon) d\varepsilon + \rho_s(eV_T) e V_m \sin(\omega_m t) + \rho_s'(eV_T) \frac{e^2 V_m^2}{2} \sin^2(\omega_m t) \tag{2.16}$$

where $\int_0^{eV_T} \rho_s(\varepsilon) d\varepsilon \propto I_T(V_T)$; $\rho_s(eV_T) \propto I_T'(V_T)$; $\rho_s'(eV_T) \propto I_T''(V_T)$.

In essence, a lock-in amplifier takes the input signal multiplied by the reference signal which is either provided from the internal oscillator or an external source. Taking into account the noise, the output of the multiplier results in:

$$\sin(\omega_m t + \varphi) \times \left(\rho_s(\epsilon) e V_m \sin(\omega_m t + \varphi_0) + \int_0^\infty a_\omega \sin(\omega_m t + \varphi_\omega) d\omega \right)$$
$$= \frac{1}{2} \rho_s(\epsilon) e V_m [\cos(\varphi - \varphi_0) + \cos(2\omega_m t + \varphi + \varphi_0)] + \ldots, \tag{2.17}$$

where a_ω and φ_ω are the amplitude and phase of the current noise at the frequency of $f = \omega/2\pi$, respectively. As it can be seen, only the frequency of the input signal is same as the reference signal can contribute to the output result depending on the phase $\varphi - \varphi_0$. All overlying current noises with broad frequencies and uncorrelated phases with respect to the reference are filtered out by the frequency low-pass. Therefore,

2.1 Scanning Tunneling Microscopy

the output of the Lock-In is directly proportional to the amplitude of the modulation bias V_m and the LDOS of sample surface.

Furthermore, the energy resolution is limited due to the modulation technique. As a matter of fact, the output dI/dV a convolution of the sample LDOS and the instrumental resolution function. The smaller modulation voltage will induce smaller broadening of the spectrum and higher energy resolution, whereas the intensity of the signal is small and thus poor signal to noise ratio. Generally, the modulation bias voltage is 3–10 mV in the STS measurement. In addition, the energy resolution is also determined by the temperature of the STM junction. Assuming a perfectly flat LDOS in the sample with only one infinite sharp peak, the temperature T broadens this peak to a Gaussian-like peak with a FWHM of 3.2 $k_B T$. Therefore, low temperature is required to obtain high resolution STS data.

2.1.3 Inelastic Electron Tunneling Spectroscopy

Most of the electrons tunnel elastically between two metal electrodes, while some tunneling electrons can lose energy by exciting the vibrations of the adsorbate, called inelastic electron tunneling process. This process takes place at a threshold bias voltage corresponding to the vibrational energy, leading to the opening of a new conductance channel. Such an effect is manifested as a slight kink in I–V curve, and a small step in dI/dV spectrum accordingly (Fig. 2.2a, b). The inelastic contribution to the current is small compared to the elastic tunneling current and is more clearly seen as a peak/dip in the second derivative of the current to the bias voltage (Fig. 2.2c), i. e. inelastic electron tunneling spectroscopy (IETS), in which the peak/dip feature denote the vibrational fingerprint.

IETS was first demonstrated in a metal-insulator-metal tunneling junction by Jaklevic and Lambe in 1966 [17]. However, the signals came from ~10^9 molecules buried at the metal-insulator interface. In 1998, Stipe et al. reported the first single-molecule IETS of acetylene on Cu (100) surface using scanning tunneling microscopy (STM-IETS), which pushed the vibrational spectroscopy down to the single bond limit [18]. The spatial distribution of the inelastic tunneling is well localized to the chemical

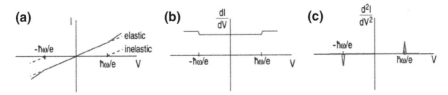

Fig. 2.2 Schematic spectra of IETS. a I–V, **b** dI/dV, and **c** d^2I/dV^2 spectra, showing the inelastic electron tunneling features at the threshold bias voltage $\hbar\omega/e$. ω is angular frequency of the vibration

bond being probed [19], allowing the creation of atomic-scale spatial images of each vibrational mode.

Soon after its birth, IETS has been extensively used to study the chemical reactions on surfaces [20–23], molecular identifications [24, 25], electron-phonon coupling [26–28], intra- and inter-molecular interaction [29–32]. With the development of molecular electronics, IETS has also been employed as a spectroscopic tool to confirm the presence of single molecule and measure the electric transport properties in different experimental setups [33–35], e.g. mechanical controllable break junction [34–36], electromigration junction [37], STM break junction [36] and so on. In addition to molecular vibration, inelastic tunneling electrons can couple to other elementary excitations, such as rotation, phonon, spin, plasmon, photon, etc., which greatly expands the scope of the IETS-related techniques.

To understand the IETS signal theoretically, one needs to study the electronic transport problem taking into account electron's interaction with molecular vibrations. Different forms of perturbative approaches, including scattering theory [13, 21, 38–40] and nonequilibrium Green's function (NEGF) method [41–50], have been used to describe the inelastic transport process. The scattering theory has the advantage of better physical transparency, and was employed in the early works modeling both the metal-insulator-metal and single molecular junction [38, 39, 51]. But its account of fermionic statistics is somewhat ad hoc. More systematic NEGF method is used to overcome this problem. Moreover, the NEGF method combined with DFT based electronic structure calculation has further advantage in the simulation of IETS signal of realistic molecular junctions.

In the NEGF method, the molecular junction is divided into three regions: the left, right electrode and the central molecule. The electrical current is expressed using the central Green's functions and self-energies. The effect of electron-vibration (e-vib) interaction is taken into account through the interaction self-energy. Assuming weak e-vib interaction, the expression for the electrical current is expanded up to the second order in the e-vib interaction matrix. Details of the theory can be found in previous publications [52, 53]. Here, we quote the result directly. The correction to the second order differential conductance due to e-vib interaction can be expressed as [52, 53]

$$\partial_{eV}^2 I = \gamma \partial_{eV}^2 I_{sym} + \lambda \partial_{eV}^2 I_{asym}. \tag{2.18}$$

To arrive at this compact expression, we mainly focus on bias range near the vibrational threshold, and have ignored the background signal that does not have vibrational feature.

Here, I_{sym} and I_{asym} are two universal functions describing the symmetric and asymmetric part of the signal

$$I_{sym} = \frac{G_0}{e} \sum_{\sigma=\pm 1} \sigma(\hbar\omega + \sigma eV)\Theta(-\sigma eV - \hbar\omega), \tag{2.19}$$

2.1 Scanning Tunneling Microscopy

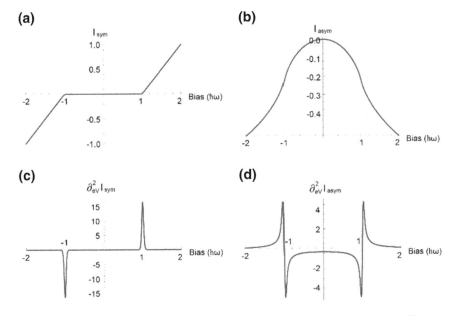

Fig. 2.3 Symmetric and asymmetric part of the vibration-induced correction to the tunneling current **a–b** and its second derivative with respect to bias **c–d**. Figure reproduced from Ref. [32]

$$I_{asym} = -\frac{G_0}{2\pi e} \sum_{\sigma=\pm 1} \sigma |\sigma\hbar\omega + eV| \ln\left|\frac{eV + \sigma\hbar\omega}{\hbar\omega}\right|. \tag{2.20}$$

And, $G_0 = 2e^2/h$ is the quantum of conductance, $\Theta(x)$ is the Heaviside step function. The two functions and their contribution to the IETS signal are plotted in Fig. 2.3.

For the Newns-Anderson-Holstein model, assuming symmetric potential drop ($\mu_L = \mu + \frac{\hbar\omega}{2}$, $\mu_R = \mu - \frac{\hbar\omega}{2}$) and $\varepsilon_0 = 0$, the expressions for the two coefficients are [39, 54]

$$\gamma = \frac{\Gamma_L \Gamma_R \Gamma^2 \left(\frac{\hbar\omega}{\Gamma}\left(\frac{\Gamma^2}{4} - \left(\mu^2 - \frac{\hbar^2\omega^2}{4}\right)\right) - 2\frac{\mu}{\Gamma}\frac{\Gamma_L - \Gamma_R}{\Gamma}\left(\frac{\Gamma^2}{4} + \left(\mu^2 - \frac{\hbar^2\omega^2}{4}\right)\right)\right)}{\left(\frac{\Gamma^2}{4} + \left(\mu - \frac{\hbar\omega}{2}\right)^2\right)^2 \left(\frac{\Gamma^2}{4} + \left(\mu + \frac{\hbar\omega}{2}\right)^2\right)^2} \tag{2.21}$$

and

$$\lambda = \frac{\Gamma_L \Gamma_R \left(\left(\mu^2 - \frac{\hbar^2\omega^2}{4}\right) - \left(\frac{\Gamma^2}{4}\right)^2 - \frac{1}{2}\Gamma^2\hbar\omega\mu\left(\frac{\Gamma_L}{\Gamma} - \frac{\Gamma_R}{\Gamma}\right)\right)}{\left(\frac{\Gamma^2}{4} + \left(\mu - \frac{\hbar\omega}{2}\right)^2\right)^2 \left(\frac{\Gamma^2}{4} + \left(\mu + \frac{\hbar\omega}{2}\right)^2\right)^2} \tag{2.22}$$

Meanwhile, in STM experiment, the molecular level couples stronger to the substrate than to the tip. Assuming $\Gamma_L \ll \Gamma_R$, we have $\mu_L = \mu, \mu_R = \mu - \hbar\omega$, the expression then simplifies to

$$\gamma \approx \frac{\Gamma_L \Gamma^2 \mu}{\left(\mu^2 + \frac{\Gamma^2}{4}\right)^2 \left((\mu - \hbar\omega)^2 + \frac{\Gamma^2}{4}\right)} \tag{2.23}$$

and

$$\lambda \approx \frac{\Gamma \Gamma_L \left(\mu^2 - \frac{\Gamma^2}{4}\right)}{\left(\mu^2 + \frac{\Gamma^2}{4}\right)^2 \left((\mu - \hbar\omega)^2 + \frac{\Gamma^2}{4}\right)} \tag{2.24}$$

This result was firstly obtained by Persson and Baratoff [39] and has been widely used to fit the experimental results.

2.1.4 Applications

The applications of STM are summarized in the following:

(1) Atomic-scale high resolution imaging

STM provides a real space view of the atomic structure of the solid surface. For instance, Au(111) surface shows a 22 × $\sqrt{3}$ reconstruction, where the atoms of surface layer occupy both the hcp and the fcc sites (Fig. 2.4). The domain walls between the hcp and the fcc regions, which are called herringbone structures, are visualized in the STM image.

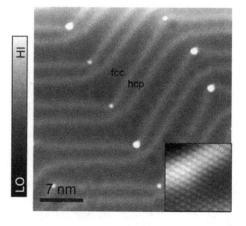

Fig. 2.4 STM image of a Au(111)- 22 × $\sqrt{3}$ reconstructed surface, showing the prominent herringbone structures. The atomically resolved STM image is inserted in the right corner

2.1 Scanning Tunneling Microscopy

(2) Probing the electronic structure of surface or adsorbate

Individual molecules adsorbed on the solid surfaces are identified in the STM images. What's more, submolecular resolution has been achieved via orbital imaging [55]. The dI/dV spectroscopy of a pentacene molecule on NaCl exhibits two distinct peak features, which are attributed to the lowest unoccupied (LUMO) and highest occupied (HOMO) molecular orbital states. Bias dependent imaging reveals the HOMO and LUMO of the individual pentacene molecules directly.

(3) Single atom/molecule manipulation

Single atoms and molecules could be manipulated in a controlled manner with STM [24, 56, 57]. The manipulation method can be either the tip-induced electric field gradient, the inelastic electron tunneling current or the mechanical force, such as pushing or pulling with the STM tip.

(4) Chemical analysis

Vibrational spectroscopy has long served as a fingerprinting technique for chemical analysis. The acetylene isotopes (C_2H_2, C_2D_2, C_2HD) on Cu(001) could be distinguished with STM-IETS, which has pushed the vibrational sensitivity down to single bond limit [18, 19, 25].

(5) Imaging the chemical structure of molecules with inelastic tunneling probe

Based on STM-IETS, Chiang et al. developed inelastic tunneling probe (itProbe) [29] to image local potential energy landscape over an adsorbed molecule via probing the vibrational frequency shift of carbon monoxide (CO) molecule on the tip. The skeletal structure and chemical bonding of the molecule are resolved in real space [29, 58].

(6) Probing elementary excitations (phonon, spin, plasmon and photon)

Phonons are the collective atomic or molecular vibrations in crystals. Gawronski et al. investigated the spatial variation of surface phonon over Au(111) surface, which arises from the site-specific probability of phonon excitation [28]. What's more, phonons in graphite were also detected with the phonon energy of 63 ± 2 meV [59].

Transitions between different spin states can be excited by exchanging spin with inelastic tunneling electrons. The cross section of spin exchanging scattering is quite large, resulting in an obvious conductance step in the first order differential conductance, which is termed as spin excitation spectroscopy. Heinrich et al. reported the first spin excitation spectra of individual manganese atoms on alumina under external magnetic field [60]. Then, various transition metal atoms, e.g. Fe, Mn, Co, on different surfaces were investigated, revealing the Landé g factor, magnetic anisotropy energy, and effective spin moment [61, 62].

Moreover, localized plasmon modes confined within the cavity between STM tip and metal substrate can be excited by inelastic electronic transitions. This tunneling electron-induced light emission was first reported in 1989 by Gimzewski et al. [63].

Ever since, the dependence of light emission properties on different parameters have been explored, such as the cavity geometry, metal surfaces, molecular adsorption [64–70]. STM-induced light emission can be further applied to study single molecular electroluminescence [66, 69, 71–75] and energy transfer at single molecular level [76, 77].

2.2 Non-contact Atomic Force Microscopy

2.2.1 *Principle*

AFM is similar to STM, except that the metal tip is replaced by a force sensor (cantilever). The cantilever is typically silicon or silicon nitride with a tip radius of curvature on the order of nanometers. When the tip is brought into proximity of a sample surface, forces between the tip and the sample lead to a deflection of the cantilever according to Hooke's law. Unlike tunneling current, which has a very short range, the forces between the tip and sample bas both short- and long-range contributions. In vacuum, there are long range van der Waals, electrostatic, and magnetic forces and short range chemical forces and Pauli repulsion force.

The van der Waals interaction is caused by fluctuations in the electric dipole moment of atoms and their mutual polarization. The van der Waals interaction potential between two hydrogen atoms is

$$U(z) = -\frac{C}{z^6} \qquad (2.25)$$

The minus sixth power dependence of the van der Waals force is valid for the interaction between two individual uncharged atoms or molecules. Due to its long-range nature and additive behavior, not only the atoms at the tip apex, but also a sizable volume of atoms in the tip and sample body contribute to the total observed van der Waals interaction. Therefore, the van der Waals interaction energy between a tip and a sample surface can be obtained by integration and is

$$U(z) = -\frac{Hr_c}{6z} \qquad (2.26)$$

where, r_c is the local radius of the tip and H is Hamaker constant, which is usually 2-3 eV for typical metals.

The chemical bond could be descried by Morse potential

$$U_{morse} = -E_{bond}\left(2e^{-\kappa(z-\sigma)} - e^{-2\kappa(z-\sigma)}\right) \qquad (2.27)$$

where E_{bond}, σ and κ are bonding energy, equilibrium distance, and a decay length, respectively. The Lennard-Jones potential is a mathematically simple model

2.2 Non-contact Atomic Force Microscopy

that approximates the interaction between a pair of neutral atoms or molecules, written as

$$U_{\text{Lennard-Jones}} = -E_{bond}\left(\left(\frac{\sigma}{z}\right)^6 - \left(\frac{\sigma}{z}\right)^{12}\right) \quad (2.28)$$

which has an attractive term $\propto z^6$ originating from the van der Waals interaction and a repulsive term $\propto z^{12}$. When the tip and sample have an electrostatic potential difference, the electrostatic interaction could not be neglected, which can be attractive or repulsive interactions between permanent charges, dipoles, quadrupoles, and in general between permanent multipoles.

2.2.2 Q-plus Sensor Based nc-AFM

Nc-AFM has two dynamic operation modes: amplitude modulation (AM) and frequency modulation (FM) modes. In vacuum, the AM mode is very slow when the Q factor is very high (~50000). So, nc-AFM works with FM operation mode. In FM-AFM (Fig. 2.5), the cantilever is driven by an actuator to oscillate at the eigenfrequency f_0 with the constant amplitude A. When the tip is approaching close to the sample surface, the tip-sample interaction will induce a change in the oscillating frequency. The frequency shift (Δf) is proportional to the tip-sample force gradient (k_{ts}) when $k_{ts} \ll k$, and k is the spring constant of the cantilever. Therefore, one can get the topography of the sample surface by measuring the frequency shift Δf.

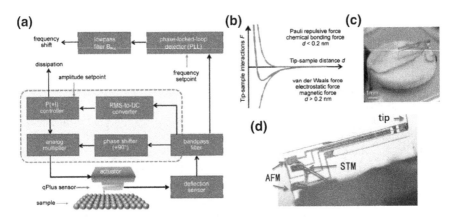

Fig. 2.5 Working principle of non-contact atomic force microscopy (NC-AFM). a Block diagram of the NC-AFM feedback loop for constant amplitude control and frequency-shift measurement. **b** Total interaction force between tip and sample. **c, d** Structure of the qPlus sensor

Fig. 2.5a display the block diagram of the FM-AFM feedback loop for constant oscillating amplitude control and frequency-shift measurement [78, 79]. The deflection signal will first enter a bandpass filter to cut off the unwanted frequency band and then split into three branches. Two of them are used to construct a feedback loop to keep the cantilever oscillating at the constant amplitude (dashed grey frame in Fig. 2.5a). The filtered defection signal flows into an rms-to-dc converter and a phase shifter. A dc signal corresponding to the rms value of the amplitude is obtained by the rms-to-dc converter and is compared with the setpoint value of the amplitude to yield the amplitude error signal. Then the amplitude error enters a proportional (P) and optical integral (I) controller, and the resulting g is multiplied with the phase shifted signal with an analog multiplier chip to drive the actuator. The phase shifter is adjusted to $\pi/2$, so that the driving signal to maintain the setpoint amplitude is minimal. Another branch signal is fed into a digital phase-locked-loop detector to yield the frequency shift signal.

A qPlus sensor is a cantilever made from a quartz tuning fork (Fig. 2.5c, d). One of the prongs is fixed to a large substrate and a tip is mounted to the other free prong. The operation of the sensor is based on the piezoelectric effect. The tip-sample interaction induces the bending of the cantilever, which causes surface charges collected by the metal electrodes. Another isolated electrode is connected to the tip to measure the tunneling current. As a result, the qPlus sensor based STM/AFM enables the simultaneous measurement of the tunneling current and force between the tip and sample. The high stiffness of the tuning fork with the spring constant $k_0 \approx 1800\,\text{N/m}$, resonance frequency $f_0 \approx 20 - 32 \times 10^3$ Hz and quality factor $Q \approx 3 - 7 \times 10^4$ allows stable operation at small oscillation amplitude (<1 Å), which is an important prerequisite for measuring short-range forces to achieve atomic resolution.

2.2.3 Applications

The applications of nc-AFM are summarized in the following:

(1) High resolution imaging

qPlus sensor based nc-AFM shows the ability to achieve superior resolution and sensitivity of single molecules/atoms, such as identifying the chemical structure and intermolecular interaction [80–84], determining the bond order [85] and chemical-reaction products [86], imaging the charge distribution within a molecule [87], measuring the force needed to move an atom [88], and even revealing the internal structure of metal clusters [89]. This is achieved by functionalizing the tip with individual atoms or molecules, such as carbon monoxide, chlorine atom as well as the recently developed copper oxide tip [90–93]. The longer-range attractive van der Waals and electrostatic forces show no atomic contrast and give rise to the dark halo surrounding the molecule. The atomic features only become visible when working at small tip heights, where the short-range Pauli repulsion between the CO molecule at the tip and the imaged molecule on the surface dominated and contribute heavily to the frequency shift.

(2) Force spectroscopy measurement with superior sensitivity

Besides imaging, force spectroscopy measurement is also very useful and can be exploited to access many different properties. To obtain the tip-sample interaction forces and energies, one need to record the frequency shift when ramping the tip-sample distance. The force can be computed by integrating frequency shift with respect to the vertical tip height using the Sader and Jarvis formula [94]. Three-dimensional force maps [95, 96] can be obtained, showing the frequency shift, force and energy maps at specific tip-sample distance [97]. Using force–distance spectroscopy, chemical contrast was demonstrated on semiconductor surfaces [98], whereby three different species of atoms could be clearly distinguished by comparing the maximum attractive force. Moreover, AFM experiments have demonstrated the capability to measure the force needed to move an atom [88] and energy required to operate the conformational switching of a single molecule [99].

(3) Probing the charge state and charge distribution

As an additional operation mode of nc-AFM, Kelvin probe force microscopy (KPFM) [8] measure the work function of surfaces, which provides information about the electronic state of the local structures on the surface [100]. KPFM can be performed by measuring the frequency shift while ramping the bias, then the local contact potential difference (LCPD) is extracted by fitting the bias spectroscopy. KPFM is extremely sensitive to charges, and experiments have demonstrated the applications in measuring and manipulating the atomic charge states [101], as well as imaging the charge distribution within an individual molecule [87]. Such information would provide valuable insight into the charge transport in molecular electronics and catalytic reactivity of adsorbates that are governed by charge state.

References

1. Binnig G, Rohrer H, Gerber C, Weibel E (1982) Tunneling through a controllable vacuum gap. Appl Phys Lett 40:178–180
2. Binning G, Rohrer H, Gerber C, Weibel E (1982) Surface studies by scanning tunneling microscopy. Phys Rev Lett 49:57–61
3. Binnig G, Rohrer H (1987) Scanning tunneling microscopy-from birth to adolescence. Rev Mod Phys 59:615–625
4. Binnig G, Rohrer H, Gerber C, Weibel E (1983) 7×7 reconstruction on Si(111) resolved in real space. Phys Rev Lett 50:120–123
5. Lang ND (1986) Theory of single-atom imaging in the scanning tunneling microscope. Phys Rev Lett 56:1164–1167
6. Binnig G, Quate CF, Gerber C (1986) Atomic force microscope. Phys Rev Lett 56:930–933
7. Kaiser WJ, Bell LD (1988) Direct investigation of subsurface interface electronic-structure by ballistic-electron-emission microscopy. Phys Rev Lett 60:1406–1409

8. Nonnenmacher M, Oboyle MP, Wickramasinghe HK (1991) Kelvin probe force microscopy. Appl Phys Lett 58:2921–2923
9. Hartmann U (1988) Magnetic force microscopy-some remarks from the micromagnetic point of view. J Appl Phys 64:1561–1564
10. Betzig E, Trautman JK, Harris TD, Weiner JS, Kostelak RL (1991) Breaking the diffraction barrier-optical microscopy on a nanometric scale. Science 251:1468–1470
11. Gao C, Wei T, Duewer F, Lu YL, Xiang XD (1997) High spatial resolution quantitative microwave impedance microscopy by a scanning tip microwave near-field microscope. Appl Phys Lett 71:1872–1874
12. Michaelson HB (1977) Work function of the elements and its periodicity. J Appl Phys 48:4729–4733
13. Bardeen J (1961) Tunnelling from a many-particle point of view. Phys Rev Lett 6:57
14. Lang ND (1986) Spectroscopy of single atoms in the scanning tunneling microscope. Phys Rev B 34:5947–5950
15. Tersoff J, Hamann DR (1983) Theory and application for the scanning tunneling microscope. Phys Rev Lett 50:1998–2001
16. Tersoff J, Hamann DR (1985) Theory of the scanning tunneling microscope. Phys Rev B 31:805–813
17. Buonsanti R, Llordes A, Aloni S, Helms BA, Milliron DJ (2011) Tunable infrared absorption and visible transparency of colloidal aluminum-doped zinc oxide nanocrystals. Nano Lett 11:4706–4710
18. Stipe BC, Rezaei MA, Ho W (1998) Single-molecule vibrational spectroscopy and microscopy. Science 280:1732–1735
19. Stipe BC, Rezaei HA, Ho W (1999) Localization of inelastic tunneling and the determination of atomic-scale structure with chemical specificity. Phys Rev Lett 82:1724–1727
20. Lee HJ, Ho W (1999) Single-bond formation and characterization with a scanning tunneling microscope. Science 286:1719–1722
21. Lauhon LJ, Ho W (2000) Control and characterization of a multistep unimolecular reaction. Phys Rev Lett 84:1527–1530
22. Kim Y, Komeda T, Kawai M (2002) Single-molecule reaction and characterization by vibrational excitation. Phys Rev Lett 89:126104
23. Komeda T, Kim Y, Kawai M, Persson BNJ, Ueba H (2002) Lateral hopping of molecules induced by excitation of internal vibration mode. Science 295:2055–2058
24. Komeda T (2005) Chemical identification and manipulation of molecules by vibrational excitation via inelastic tunneling process with scanning tunneling microscopy. Prog Surf Sci 78:41–85
25. Ho W (2002) Single-molecule chemistry. J Chem Phys 117:11033–11061
26. Grobis M et al (2005) Spatially dependent inelastic tunneling in a single metallofullerene. Phys Rev Lett 94:136802
27. Lee J et al (2006) Interplay of electron-lattice interactions and superconductivity in $Bi_2Sr_2CaCu_2O_{8+\delta}$. Nature 442:546–550
28. Gawronski H, Mehlhorn M, Morgenstern K (2008) Imaging phonon excitation with atomic resolution. Science 319:930–933
29. Chiang CL, Xu C, Han ZM, Ho W (2014) Real-space imaging of molecular structure and chemical bonding by single-molecule inelastic tunneling probe. Science 344:885–888
30. Li SW et al (2015) Rotational spectromicroscopy: imaging the orbital interaction between molecular hydrogen and an adsorbed molecule. Phys Rev Lett 114:206101
31. Han Z et al (2017) Probing intermolecular coupled vibrations between two molecules. Phys Rev Lett 118:036801
32. Guo J et al (2016) Nuclear quantum effects of hydrogen bonds probed by tip-enhanced inelastic electron tunneling. Science 352:321–325
33. Beebe JM, Moore HJ, Lee TR, Kushmerick JG (2007) Vibronic coupling in semifluorinated alkanethiol junctions: Implications for selection rules in inelastic electron tunneling spectroscopy. Nano Lett 7:1364–1368

34. Taniguchi M, Tsutsui M, Yokota K, Kawai T (2009) Inelastic electron tunneling spectroscopy of single-molecule junctions using a mechanically controllable break junction. Nanotechnology 20:434008
35. Kim Y et al (2011) Conductance and vibrational states of single-molecule junctions controlled by mechanical stretching and material variation. Phys Rev Lett 106:196804
36. Bruot C, Hihath J, Tao NJ (2012) Mechanically controlled molecular orbital alignment in single molecule junctions. Nat Nanotech 7:35–40
37. Song H et al (2009) Vibrational spectra of metal-molecule-metal junctions in electromigrated nanogap electrodes by inelastic electron tunneling. Appl Phys Lett 94:103110
38. Scalapino DJ, Marcus SM (1967) Theory of inelastic electron-molecule interactions in tunnel junctions. Phys Rev Lett 18:459
39. Persson BNJ, Baratoff A (1987) Inelastic electron tunneling from a metal tip: the contribution from resonant processes. Phys Rev Lett 59:339–342
40. Lorente N, Persson M, Lauhon LJ, Ho W (2001) Symmetry selection rules for vibrationally inelastic tunneling. Phys Rev Lett 86:2593–2596
41. Galperin M, Ratner MA, Nitzan A (2004) Inelastic electron tunneling spectroscopy in molecular junctions: peaks and dips. J Chem Phys 121:11965–11979
42. Pecchia A et al (2004) Incoherent electron-phonon scattering in octanethiols. Nano Lett 4:2109–2114
43. Chikkannanavar SB, Luzzi DE, Paulson S, Johnson AT (2005) Synthesis of peapods using substrate-grown SWNTs and DWNTs: an enabling step toward peapod devices. Nano Lett 5:151–155
44. Sergueev N, Roubtsov D, Guo H (2005) Ab initio analysis of electron-phonon coupling in molecular devices. Phys Rev Lett 95:146803
45. Viljas JK, Cuevas JC, Pauly F, Hafner M (2005) Electron-vibration interaction in transport through atomic gold wires. Phys Rev B 72:245415
46. Solomon GC et al (2006) Understanding the inelastic electron-tunneling spectra of alkanedithiols on gold. J Chem Phys 124:094704
47. Frederiksen T, Paulsson M, Brandbyge M, Jauho AP (2007) Inelastic transport theory from first principles: methodology and application to nanoscale devices. Phys Rev B 75:205413
48. Galperin M, Ratner MA, Nitzan A (2007) Molecular transport junctions: vibrational effects. J Phys Condens Matter 19:103201
49. Paulsson M, Frederiksen T, Brandbyge M (2005) Modeling inelastic phonon scattering in atomic- and molecular-wire junctions. Phys. Rev. B 72:201101
50. Chen YC, Zwolak M, Di Ventra M (2003) Local heating in nanoscale conductors. Nano Lett. 3:1691–1694
51. Lambe J, Jaklevic RC (1968) Molecular vibration spectra by inelastic electron tunneling. Phys Rev 165:821
52. Paulsson M, Frederiksen T, Brandbyge M (2005) Modeling inelastic phonon scattering in atomic- and molecular-wire junctions. Phys Rev B 72:201101
53. Lü JT et al (2014) Efficient calculation of inelastic vibration signals in electron transport: beyond the wide-band approximation. Phys Rev B 89:081405
54. Egger R, Gogolin AO (2008) Vibration-induced correction to the current through a single molecule. Phys Rev B 77:113405
55. Repp J, Meyer G, Stojkovic SM, Gourdon A, Joachim C (2005) Molecules on insulating films: scanning-tunneling microscopy imaging of individual molecular orbitals. Phys Rev Lett 94:026803
56. Stroscio JA, Eigler DM (1991) Atomic and molecular manipulation with the scanning tunneling microscope. Science 254:1319–1326
57. Stroscio JA, Celotta RJ (2004) Controlling the dynamics of a single atom in lateral atom manipulation. Science 306:242–247
58. Han ZM et al (2017) Imaging the halogen bond in self-assembled halogenbenzenes on silver. Science 358:206+

59. Zhang YB et al (2008) Giant phonon-induced conductance in scanning tunnelling spectroscopy of gate-tunable graphene. Nat Phys 4:627–630
60. Heinrich AJ, Gupta JA, Lutz CP, Eigler DM (2004) Single-atom spin-flip spectroscopy. Science 306:466–469
61. Hirjibehedin CF et al (2007) Large magnetic anisotropy of a single atomic spin embedded in a surface molecular network. Science 317:1199–1203
62. Donati F et al (2013) Magnetic moment and anisotropy of individual Co atoms on graphene. Phys Rev Lett 111:236801
63. Gimzewski JK, Sass JK, Schlitter RR, Schott J (1989) Enhanced photon-emission in scanning tunnelling microscopy. Europhys. Lett. 8:435–440
64. Berndt R et al (1995) Atomic-resolution in photon-emission induced by a scanning tunneling microscope. Phys Rev Lett 74:102–105
65. Nazin GV, Qiu XH, Ho W (2003) Atomic engineering of photon emission with a scanning tunneling microscope. Phys Rev Lett 90:216110
66. Qiu XH, Nazin GV, Ho W (2003) Vibrationally resolved fluorescence excited with submolecular precision. Science 299:542–546
67. Berndt R et al (1993) Photon-emission at molecular resolution induced by a scanning tunneling microscope. Science 262:1425–1427
68. Wang T, Boer-Duchemin E, Zhang Y, Comtet G, Dujardin G (2011) Excitation of propagating surface plasmons with a scanning tunnelling microscope. Nanotechnology 22:175201
69. Lutz T et al (2013) Molecular orbital gates for plasmon excitation. Nano Lett. 13:2846–2850
70. Kuhnke K, Große C, Merino P, Kern K (2017) Atomic-scale imaging and spectroscopy of electroluminescence at molecular interfaces. Chem Rev 117:5174
71. Dong ZC et al (2010) Generation of molecular hot electroluminescence by resonant nanocavity plasmons. Nat Photon 4:50–54
72. Schneider NL, Lu JT, Brandbyge M, Berndt R (2012) Light emission probing quantum shot noise and charge fluctuations at a biased molecular junction. Phys Rev Lett 109:186601
73. Dong ZC et al (2004) Vibrationally resolved fluorescence from organic molecules near metal surfaces in a scanning tunneling microscope. Phys Rev Lett 92:086801
74. Reecht G et al (2014) Electroluminescence of a polythiophene molecular wire suspended between a metallic surface and the tip of a scanning tunneling microscope. Phys. Rev. Lett. 112:047403
75. Chong MC et al (2016) Narrow-line single-molecule transducer between electronic circuits and surface plasmons. Phys Rev Lett 116:036802
76. Zhang Y et al (2016) Visualizing coherent intermolecular dipole-dipole coupling in real space. Nature 531:623
77. Imada H et al (2016) Real-space investigation of energy transfer in heterogeneous molecular dimers. Nature 538:364
78. Giessibl FJ (2003) Advances in atomic force microscopy. Rev Mod Phys 75:949–983
79. Liu M-X, Li S-C, Zha Z-Q, Qiu X-H (2017) Research progress and applications of qPlus noncontact atomic force microscopy. Acta Phy Chim Si 33:183–197
80. Gross L, Mohn F, Moll N, Liljeroth P, Meyer G (2009) The chemical structure of a molecule resolved by atomic force microscopy. Science 325:1110–1114
81. Albrecht F, Neu M, Quest C, Swart I, Repp J (2013) Formation and characterization of a molecule-metal-molecule bridge in real space. J Am Chem Soc 135:9200–9203
82. Zhang J et al (2013) Real-space identification of intermolecular bonding with atomic force microscopy. Science 342:611–614
83. Kawai S et al (2016) Van der Waals interactions and the limits of isolated atom models at interfaces. Nat Commun 7:11559
84. Hämäläinen SK et al (2014) Intermolecular contrast in atomic force microscopy images without intermolecular bonds. Phys Rev Lett 113:186102
85. Gross L et al (2012) Bond-order discrimination by atomic force microscopy. Science 337:1326–1329

References

86. de Oteyza DG et al (2013) Direct imaging of covalent bond structure in single-molecule chemical reactions. Science 340:1434–1437
87. Mohn F, Gross L, Moll N, Meyer G (2012) Imaging the charge distribution within a single molecule. Nat Nanotech 7:227–231
88. Ternes M, Lutz CP, Hirjibehedin CF, Giessibl FJ, Heinrich AJ (2008) The force needed to move an atom on a surface. Science 319:1066–1069
89. Emmrich M et al (2015) Subatomic resolution force microscopy reveals internal structure and adsorption sites of small iron clusters. Science 348:308–311
90. Bamidele J et al (2012) Chemical tip fingerprinting in scanning probe microscopy of an oxidized Cu(110) surface. Phys Rev B 86:155411
91. Monig H et al (2013) Understanding scanning tunneling microscopy contrast mechanisms on metal oxides: a case study. Acs Nano 7:10233–10244
92. Monig H et al (2016) Submolecular imaging by noncontact atomic force microscopy with an oxygen atom rigidly connected to a metallic probe. ACS Nano 10:1201–1209
93. Monig H et al (2018) Quantitative assessment of intermolecular interactions by atomic force microscopy imaging using copper oxide tips. Nat Nanotech 13:371+
94. Sader JE, Sugimoto Y (2010) Accurate formula for conversion of tunneling current in dynamic atomic force spectroscopy. Appl Phys Lett 97:043502
95. Albers BJ et al (2009) Three-dimensional imaging of short-range chemical forces with picometre resolution. Nat Nanotech 4:307–310
96. Baykara MZ, Schwendemann TC, Altman EI, Schwarz UD (2010) Three-dimensional atomic force microscopy-taking surface imaging to the next level. Adv Mater 22:2838–2853
97. Moll N, Gross L, Mohn F, Curioni A, Meyer G (2010) The mechanisms underlying the enhanced resolution of atomic force microscopy with functionalized tips. New J Phys 12:125020
98. Sugimoto Y et al (2007) Chemical identification of individual surface atoms by atomic force microscopy. Nature 446:64–67
99. Loppacher C et al (2003) Direct determination of the energy required to operate a single molecule switch. Phys Rev Lett 90:066107
100. Melitz W, Shen J, Kummel AC, Lee S (2011) Kelvin probe force microscopy and its application. Surf Sci Rep 66:1–27
101. Gross L et al (2009) Measuring the charge state of an adatom with noncontact atomic force microscopy. Science 324:1428–1431

Chapter 3
Submolecular-Resolution Imaging of Interfacial Water

3.1 Introduction

Water-solid interactions is ubiquitous in nature and play a key role in numerous scientific and technological fields, such as photocatalytic water splitting, heterogeneous and homogeneous catalysis, electrochemistry, corrosion and lubrication [1–8]. Characterization of H-bonded networks formed on surfaces has been one of the fundamental and key issues in water science. Despite the massive STM studies of interfacial water we summarized in Chap. 1, it is still a great challenge to resolve the internal structure of water molecule, that is, the directionality of OH.

The main difficulties lie in the following four aspects. First, the water molecules are interconnected by H bonds, which are much weaker than the covalent and ionic bonds. It is highly possible that the water molecules can be disturbed by the STM tip during the imaging process and deviate from its original structure. Second, due to the close shell nature of water molecule, the frontier orbitals of water are located very far away from the Fermi level (E_F). The electrons from the STM tip have little chance to tunnel into the molecular resonance of water, leading to a very poor signal-to-noise ratio. Third, H atoms are very small and light, so it is very hard to access the internal degree of freedom of water. Finally, the atomic resolution of organic molecules with nc-AFM is typically achieved at the very small tip-molecule separation where the short-range Pauli repulsion force is dominant [9–13], while the water structure may be easily disturbed at such small tip heights.

In this chapter, we report the submolecular-resolution imaging of water molecules adsorbed on a Au-supported NaCl(001) film with STM [14] and nc-AFM [15]. In the STM experiments, we first decouple electronically the water molecule from the metal substrate by inserting an insulating NaCl layer and then employed the STM tip as a top gate to tune controllably the molecular density of states of water around the Fermi level (E_F). These key steps enable the direct visualizing of frontier molecular orbitals of adsorbed water, which allows discriminating the orientation of the monomers, the H-bond directionality of the tetramers in real space and characterization of H-bonded

© Springer Nature Singapore Pte Ltd. 2018
J. Guo, *High Resolution Imaging, Spectroscopy and Nuclear Quantum Effects of Interfacial Water*, Springer Theses, https://doi.org/10.1007/978-981-13-1663-0_3

water clusters and overlayers on NaCl(001) film [14, 16]. In addition, we also achieve the submolecular-resolution imaging of water nanoclusters on the NaCl(001) film by probing the high-order electrostatic force using a qPlus-based nc-AFM with a CO-terminated tip [15]. The non-invasive AFM imaging technique may open a new avenue of studying the intrinsic structure and dynamics of ice or water on surfaces, ion hydration and biological water with atomic precision.

3.2 Methods

3.2.1 STM/AFM Experiments

The experiments were performed with a combined nc-AFM/STM system (Createc, Germany) at 5 K with the base pressure better than 8×10^{-11} torr using a qPlus sensor equipped with a W tip (spring constant $k_0 \approx 1800$ N m^{-1}, resonance frequency $f_0 = 23.7$ kHz, and quality factor $Q \approx 50,000$). The Au(111) single crystal was repeatedly sputtered with Argon and annealed at about 900 K, until a clean Au(111)-22× $\sqrt{3}$ reconstructed surface was obtained. The NaCl (Sigma Aldrich, 99.999%) was then evaporated thermally from a Knudsen-cell at a temperature of 720 K onto the Au(111) surface held at 275 K. The ultrapure H_2O (Sigma Aldrich, deuterium-depleted) was further purified under vacuum by freeze-thaw cycles to remove remaining impurities. The H_2O molecules were dosed in situ onto the sample surface at 5 and 77 K depending on the coverage and specific size of the water molecules, through a dosing tube, which pointed toward the sample from a distance of about 6 cm.

The STM measurements were performed at 5 or 77 K with electrochemically etched tungsten tips, which were sharpened and optimized in situ by controlled field-emission and tip-cash procedures. Bias voltage refers to the sample voltage with respect to the tip. All the STM topographic images were obtained in the constant current mode. The scanning tunneling spectroscopy (STS) dI/dV spectra were acquired using lock-in detection of the tunneling current by adding a 10 m V_{rms} modulation at 250 Hz to the sample bias. STM and AFM image processing was performed by *Nanotec* WSxM [18].

All of the AFM (frequency shift, Δf) images were obtained in constant-height mode at 5 K with Cl- or CO-terminated tips. The Cl-terminated tip was obtained by picking up a Cl atom from the NaCl surface by approaching the tip in close with the Cl atom (set point: V = 5 mV and I = 5 nA), which can be confirmed by improved resolution and the appearance of a single Cl vacancy by scanning the same area (Fig. 3.1). The CO-tip was obtained by positioning the tip over a CO molecule adsorbed on the NaCl film at a set point of 100 mV and 20 pA, followed by increasing the bias voltage to 200 mV. The controllable manipulation of water monomers to construct water tetramers was achieved with the Cl-terminated tip at

3.2 Methods

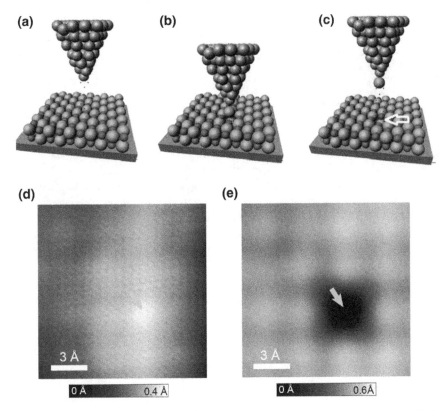

Fig. 3.1 Preparation of a Cl-functionalized tip. a–c Schematic procedure to prepare a Cl-tip. Approaching a bare sharp STM tip towards the Cl atom of NaCl surface in close proximity. **b** Until a single Cl atom hops onto the apex of the STM tip (**c**). **d, e** STM images of the same NaCl(001) surface acquired with a bare tip (**d**) and Cl-tip (**e**), respectively. The atomic resolution arising from the Cl^- anions. The cyan arrow denotes the position where the STM tip is positioned. Set point: **d** 50 mV, 100 pA; **e** 50 mV, 50 pA. The STM images were obtained at 5 K. Reproduced with permission from [17]

the set point: V = 10 mV, I = 150 pA. The contrast of the images does not change at different scanning directions, so we did not explicitly denote the scanning directions unless it is necessary.

3.2.2 DFT Calculations

DFT calculations were performed using Vienna Ab initio Simulation Package code (VASP) [19, 20] to study the adsorption configuration of water monomers and nanoclusters on the NaCl surface. Van der Waals corrections for dispersion forces were considered within the vdW-DF scheme by using the optB88-vdW method [21]. Projector augmented-wave pseudopotentials [22] were used together with a plane wave basis set and a cutoff energy of 550 eV for the expansion of the electronic wave functions. In the DFT calculation, we constructed a NaCl/Au coincidence structure by superposing a NaCl (2 × 2) unit cell on a $\begin{pmatrix} 3 & 1 \\ 1 & 3 \end{pmatrix}$ superstructure of the Au(111) substrate with a residual strain of about 5%. To match the square symmetry of the NaCl(001) lattice, the angle of the $\begin{pmatrix} 3 & 1 \\ 1 & 3 \end{pmatrix}$ supercell of the Au(111) substrate was changed from 82° to 90°. The Au substrate was modelled by a four-layer slab and a bilayer NaCl(001) slab with a lattice constant of 5.665 Å was used.

In the calculation of adsorption of water clusters, water/NaCl(001) structures were built with large unit cells to keep the error of water-image interaction at a negligible level. The geometry optimizations were run with the energy criterion of 5 × 10^{-5} eV and the adsorption energy was calculated by subtracting the total energy of the nH$_2$O/NaCl(001) structure from the sum of the energies of the relaxed bare NaCl(001) substrate and n isolated water molecules in gas phase:

$$E_{ads} = E[(NaCl(001))_{relaxed}] + n \times E[(H_2O)_{gas}] - E[(NaCl(001) + nH_2O)_{relaxed}]$$

Energy barrier of water dimer was determined using the climbing image nudged elastic band (cNEB) method [23] with the force criterion of 0.02 eV Å^{-1}. The adsorption of 2D ice was calculated using a 3 × 3 periodic tetramer-based bilayer ice slab on a 6 × 6 NaCl(001) substrate. Dozens of initial structures were generated with random distribution of tetramer chirality and Bjerrum-defect orientations.

3.2.3 AFM Simulations

The Δf images were simulated with a molecular mechanics model including the electrostatic force, based on the methods described in Refs. [24, 25]. We used the following parameters of the flexible probe-particle tip model: the effective lateral stiffness $k = 0.5$ N m^{-1} and effective atomic radius $R_c = 1.66$ Å. In order to extract the effect of electrostatics more clearly and to make z-distance directly comparable, we used the same stiffness and atomic radius to simulate AFM images acquired with CO and Cl-terminated tips. Noteworthy, the simulated Δf images using different atomic radius of the probe particle to mimick CO ($R_c = 1.66$ Å) and Cl ($R_c = 1.95$ Å) tip-apex models with the same effective charges display essentially the same features. The input electrostatic potentials of water tetramer and other water clusters were obtained by DFT calculation using the VASP code with a plane-wave cutoff 600 and 550 eV, respectively.

3.3 Submolecular-Resolution Imaging of Interfacial Water with STM

3.3.1 Orbital Imaging of Water Monomers

The clean Au(111) shows the characteristic 22 × $\sqrt{3}$ reconstruction surface with the herringbone structures, which could be resolved clearly with STM (Fig. 2.4). Bilayer NaCl islands with (001)-termination formed at the step edges of the Au(111) substrate with a carpet-like growth mode (Fig. 3. 2a). In the atomically resolved STM image of the NaCl film (Fig. 3. 2b), the Cl$^-$ anions are visualized as round protrusions and the Na$^+$ cations are invisible (Fig. 3. 2b) because of the negligible DOS around the E_F at the Na$^+$ sites [26]. To obtain isolated water monomers on NaCl(001) film (Fig. 3. 2d), we dose water molecules at 5 K to freeze the molecule and prevent the diffusion and clustering.

When scanning with a sharp tip, the STM images of water monomer exhibited symmetric double-lobe structure with two orthogonal adsorption orientations (Fig. 3.2e, f). With the square lattice of the underling NaCl surface overlaid, we found that the water monomer adsorbed on the top site of Na$^+$ along the orientation of Na$^+$–Cl$^-$ direction. DFT calculations revealed that water monomer adsorbed on the NaCl surface with a "standing" configuration with one OH points upward to the vacuum, the other OH points toward the NaCl surface (Fig. 3.2g, h). The "standing" structure is against the previous reported "flat" configuration [27–29] due to the long-range van der Waals forces from the Au substrate.

The double-lobe structure of water monomer is typically obtained at positive bias (Fig. 3.3a), which agrees perfectly with the calculated HOMO of the water monomer (Fig. 3.3d). When switching to negative bias, the double-lobe structure almost fades away and an egg-shaped lobe emerges within the nodal plane (Fig. 3.3b), closely resembling the calculated LUMO feature of the water monomer (Fig. 3.3e). We notice that there still exists a tiny contribution from the HOMO in Fig. 3.3b, which appears as two faint ring structures (denoted by arrows in Fig. 3.3b). More interestingly, the HOMO and LUMO can be imaged simultaneously and the relative contributions from HOMO and LUMO could be controlled by choosing an appropriate bias voltage (Fig. 3.3c). It is worth noting that such a composite STM image is not a simple superposition of the HOMO and LUMO (Fig. 3.3f), but probably arises from the interference effect between the tunneling paths through the two orbitals.

Since molecular orbitals are spatially locked together with the geometric structures, the submolecular orbital imaging of water monomer shows the possibility of distinguishing the orientation of the water molecule on the NaCl surface in real space. First, in the HOMO image, the mirror symmetry of the two lobes with respect to the nodal plane (Fig. 3.3a) indicates that the HOH plane of water is perpendicular to the surface. Second, the egg-shaped LUMO lobe (Fig. 3.3b) with only one axis of symmetry, allows us to determine that the flat OH bond of the monomer is oriented along the [010] direction of the NaCl(001) surface. We also noticed that the LUMO feature does not develop from the center but towards the lower edge of the HOMO.

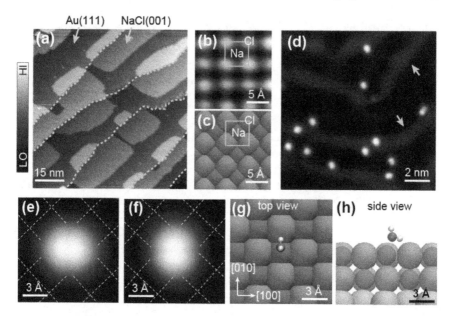

Fig. 3.2 Water monomers adsorbed on NaCl(001)/Au(111). a STM topography of bilayer NaCl(001) islands grown on the Au(111) surface. Step edges of the Au(111) surface are highlighted by skyblue dotted lines. Set point: V = 2.0 V and I = 9 pA. **b, c** Atomically resolved STM image and molecular model of NaCl(001) surface. Protrusions denote anions Cl$^-$. Set point: V = 50 mV and I = 20 pA. The unit-cells are highlighted by green squares in (**b**) and (**c**). **d** Zoom-in STM topography of a NaCl(001) island after dosing about 0.01 bilayer (BL) water molecules at 5 K. The single protrusions correspond to individual water monomers. Herringbone structures of the underlying Au(111) substrate are denoted by skyblue arrows. Set point: V = 100 mV and I = 50 pA. **e, f** STM images of two water monomers with orthogonal orientations. Square lattices of the NaCl(001) surface arising from Cl$^-$ are depicted by white grids. Set point: V = 100 mV and I = 50 pA. **g, h**, Top and side views, respectively, of the calculated adsorption configuration of a water monomer. O, H, Au, Cl$^-$ and Na$^+$ are denoted by red, white, golden, grey and dark-cyan spheres, respectively. Reproduced with permission from [14]

This feature implicates that the directionality of the upright OH of the monomer is not strictly vertical but slightly tilts away from the surface normal towards the [0$\bar{1}$0] direction. Those observations are in excellent agreement with the DFT calculation results (Fig. 3.2g, h).

In addition to the symmetric double-lobe structure, asymmetric water monomers have been observed as well because of the inhomogeneity of the NaCl surface resulting from the herringbone structures of Au substrate. Figure 3.4 shows the orbital imaging of such an asymmetric monomer. Similar to the symmetric monomer, the HOMO of the asymmetric monomer gradually evolves into the LUMO when the polarity of the bias voltage changes from positive to negative (Fig. 3.4a–d). However, the absence of the mirror symmetry with respect to the nodal plane of the HOMO (Fig. 3.4a) suggests that the HOH plane of water is no longer perpendicular to the surface but tilts towards the [$\bar{1}$00] direction of the NaCl(001) surface (Fig. 3.4e, g).

3.3 Submolecular-Resolution Imaging of Interfacial Water with STM

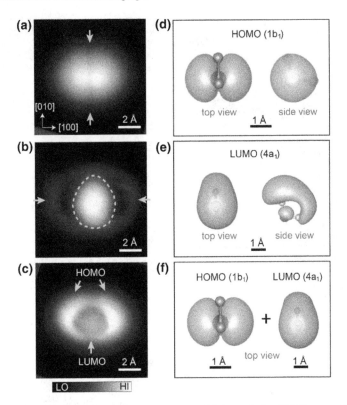

Fig. 3.3 Orbital imaging of a water monomer. a–c High-resolution STM images of HOMO (**a**), LUMO (**b**), HOMO + LUMO (**c**) of a water monomer. The nodal plane of the HOMO is highlighted by the cyan arrows in (**a**). The dashed oval and arrows in (**b**) denote the egg shape of the LUMO and the faint ring structures arising from the remaining HOMO, respectively. Set point **a** 100 mV, 500 pA; **b** −100 mV, 800 pA; **c** −50 mV, 550 pA. **d–f** Calculated HOMO, LUMO, HOMO + LUMO of a water monomer adsorbed on NaCl(001) by plotting isosurfaces of charge densities integrated over 1 eV of the HOMO/LUMO tail close to E_F. O and H atoms of H_2O are denoted by red and light-gray spheres, respectively. The STM images were obtained at 5 K. Reproduced with permission from [14]

The tilt of the HOH plane can be also evidenced in the LUMO image (Fig. 3.4d), where the LUMO lobe does not align with the [010] direction in clear contrast to the case of the symmetric monomer. In addition, the orientation of the LUMO lobe with respect to the [010] direction allows us to determine that the flat OH bond of the asymmetric monomer is oriented along the $[0\bar{1}0]$ direction of the NaCl(001) surface (Fig. 3.4e, h). By carefully comparing the STM orbital images (Fig. 3.4a, d) with the calculated isosurfaces of HOMO and LUMO (Fig. 3.4g, h), one can readily discriminate the bond orientation of different asymmetric monomers.

When tuning the bias voltage and the tunneling current in a systematic way (Fig. 3.5), several general features can be extracted: (1) the HOMO and LUMO features only emerge at small tip heights and become prominent as the tip height

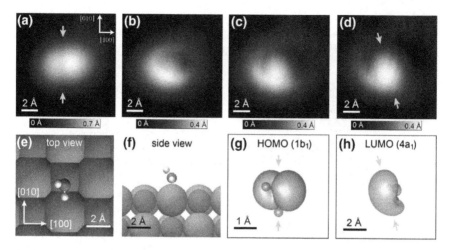

Fig. 3.4 Orbital imaging of an asymmetric water monomer adsorbed on the surface. a–d High-resolution STM images of HOMO, HOMO + LUMO, LUMO of a water monomer. Set point: **a** 10 mV, 100 pA; **b** −200 mV, 100 pA; **c** −250 mV, 100 pA; **d** −300 mV, 100 pA. **e, f** Top and side views, respectively, of the calculated adsorption configuration of an asymmetric water monomer. **g, h** Calculated isosurfaces of the charge density of HOMO and LUMO, respectively. The STM images were obtained at 5 K. Reproduced with permission from Ref. [14]

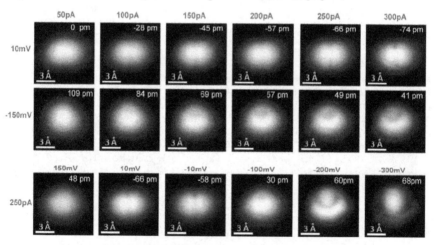

Fig. 3.5 Orbital imaging of a symmetric water monomer as functions of the bias (**bottom row**) and the tunneling current (**upper two rows**). The relative tip height (z) at the center of the monomer, referenced to the gap set with: V = 10 mV and I = 50 pA, is inserted in the upper right corner of each image. Reproduced with permission from [14]

decreases; (2) the HOMO gradually evolves into the LUMO when the polarity of the bias voltage changes from positive to negative; (3) the images acquired around zero bias voltages are dominantly contributed by the HOMO, whereas the LUMO becomes overwhelming when the bias voltage drops below −100 mV.

3.3.2 The Mechanism of Orbital Imaging

In order to explore the mechanism of orbital imaging of water, we calculated the projected density of states (PDOS) for the water monomer adsorbed on the Au-supported NaCl(001) films (Fig. 3.6a). Due to the decoupling effect of NaCl bilayer (Fig. 3.6b), the native molecular orbitals of water are mostly preserved and the broadenings of the HOMO and LUMO are very small, resulting in negligible DOS within the energy range of ±2 V. In principle, according to the previously demonstrated orbital imaging of organic molecules on NaCl surface [30], one can directly image the HOMO and LUMO through resonant tunneling. However, this method is not applicable to water molecules, since the HOMO and LUMO locate far away from E_F (Fig. 3.6a) and the water would become unstable under such large bias voltages because of vibrational excitation induced by high-energy inelastic tunneling electrons [31, 32]. In such a case, when scanning the water monomer at small bias voltage, electrons from the tip should tunnel directly into the Au(111) substrate and no orbital structures of water can be probed, unless there exist other mechanisms bringing about the development of HOMO and LUMO states near E_F.

It is known that the STM tip not only acts as a probe, but may electronically couple with the water molecule in agreement with the Newns-Anderson model [33], resulting in the enhancement of molecular DOS around E_F [34]. Then we calculated PDOS (Fig. 3.6c) of a water monomer with a small tip-water separation (3 Å) to

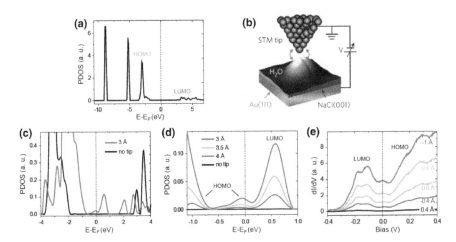

Fig. 3.6 Gating the HOMO and LUMO states of a water monomer towards the E_F via tip-water coupling. **a** PDOS of a water monomer adsorbed on NaCl(001) projected onto the water molecule. **b** Schematic of the experimental set-up. **c** Calculated PDOS of a water monomer with and without tip-water coupling. **d** Zoom in PDOS of water monomer at different tip heights. The tip height is defined as the vertical distance between the endmost atom of the STM tip and the oxygen atom of water. **e** dI/dV spectra of water monomer measured at different tip-water separations with the NaCl background signal removed. Z offsets highlighted in (**e**) are referenced to the gap set with: V = 100 mV and I = 50 pA. Reproduced from Ref. [14] with permission

include the coupling between the tip and molecular states, which exhibits considerable broadenings and shifts of the HOMO and LUMO towards E_F. To explore the coupling effect from the tip in a systematic way, we show the zoom-in PDOS near E_F with the tip-water separation changing from 3 to 4 Å (Fig. 3.6d). It is very obvious that the molecular DOS is already quite prominent near the E_F at a tip-water separation of 4 Å, in which the tail states below −0.5 eV and the states around E_F are mostly of HOMO character, whereas the states around 0.6 eV are mostly of LUMO character (Fig. 3.6d). With decreasing tip height, both the HOMO- and LUMO-like states near the E_F are enhanced owing to the increased tip-molecule coupling.

Experimentally, we could measure the dI/dV spectra at different tip heights (Fig. 3.6e) to compare with the calculated results. It is worth to mention that the background signals arising from the direct tunneling between the tip and the Au substrate were removed from the original dI/dV spectra with the method we developed in Ref. [14]. The features in dI/dV spectra are in qualitative agreement with the calculated PDOS. On one hand, a tail feature developed from the positive bias and a broad peak appeared around −0.15 V, which are, respectively, attributed to the HOMO and LUMO states from the bias-dependent orbital images we shown in Fig. 3.3. On the other hand, the magnitudes of those states increase as the tip approaches the monomer.

In spite of the highly agreement of the main feature of the spectra, there exist two discrepancies. First, the energy scale between the dI/dV spectra and the PDOS is not the same, which may result from the uncertainty of the tip apex. In the STM experiment, we also noticed that the orbital imaging is very sensitive to the details of the tip apex (Fig. 3.7). The tip in Fig. 3.7a is sensitive both to the LUMO and HOMO states. In contrast, tips in Fig. 3.7b, c selectively enhance the LUMO and HOMO states within the bias range investigated, respectively. However, different STM tips yield similar HOMO and LUMO features in spite of the variation in the energy scale. Second, for the STM setup shown in Fig. 3.6b where the bias voltage is applied to the sample, one would expect to probe the HOMO at negative biases and the LUMO at positive biases, which is against the experimental observations (Fig. 3.3). This is rationalized by the case in our experiment, where tip-water coupling is significantly stronger than the water-substrate (Au) coupling, such that the molecular states of the adsorbed water are pinned to the E_F of the STM tip instead of the Au substrate when the bias is applied. Thus, the role of the tip and sampler are inversed. The tip becomes a "nano-substrate" and the Au substrate could be considered as a macroscopic "tip", leading to observations of the HOMO at positive biases and the LUMO at negative biases.

Overall, the submolecular orbital imaging of water monomers on NaCl surface is achieved because of following key points. First, the water molecules are decoupled electronically from the metal substrate by inserting an insulating NaCl layer. So, the native molecular orbitals of water are mostly preserved. Second, the STM tip is employed as a top gate to tune controllably the molecular DOS of water around the Fermi level through the tip-water coupling. Then the HOMO and LUMO are shifted and broadened towards E_F, thus enabled orbital imaging of water molecules at around E_F.

3.3 Submolecular-Resolution Imaging of Interfacial Water with STM   53

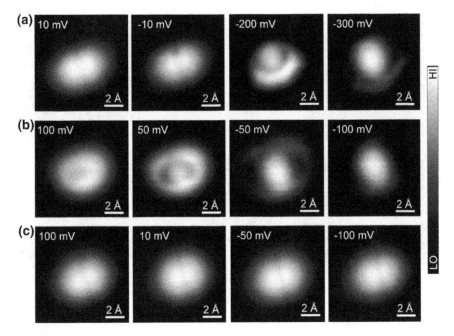

Fig. 3.7 Orbital imaging of different water monomers obtained with three different types of tip apex. The bias is inserted in the images and the tunneling current is 250 pA (**a**), 50 pA (**b** and **c**). Different tips show the selectively sensitivity to HOMO and LUMO. The STM images were obtained at 5 K. Reproduced from Ref. [14] with permission

3.3.3 Discrimination of H-Bond Directionality of Water Tetramers

The water molecule has very low mobility on the NaCl(001) surface at 5 K, such that only isolated water monomers can be found. The formation of tetramers is thus kinetically forbidden. Using a Cl-functionalized STM tip, individual water monomers could be manipulated in a well-controlled manner due to the electrostatic interaction between the Cl-tip and the dipole water molecule. This provides an efficient way to construct water tetramers and other clusters. Figure 3.8 shows a complete manipulation sequence of assembling the monomers (Fig. 3.8a) to form a dimer (Fig. 3.8c), a trimer (Fig. 3.8d), and a tetramer (Fig. 3.8e) by "pulling" the water monomer along the predesigned trajectories as highlighted by blue dashed arrows in Fig. 3.8. We note that the water dimer and trimer are not very stable such that they can be easily disturbed by the tip during scanning. However, the tetramer is quite stable once formed, allowing long-term imaging and spectroscopic measurements . It is worth

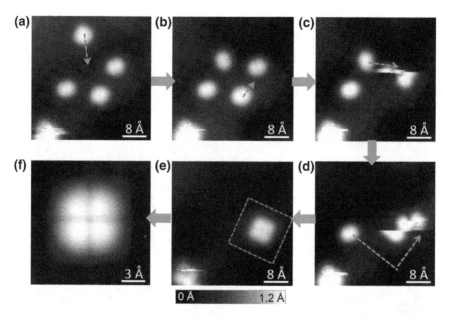

Fig. 3.8 Construction a water tetramer with the Cl-terminated tip. Water monomers were manipulated by the tip along the trajectories highlighted by the blue dashed arrows in (**a–d**). Set point of (**a–e**): 80 mV, 50 pA. **e** The formation of a water tetramer. **f** Zoom-in STM image of the constructed water tetramer. Set point: 20 mV, 80 pA, 5 K. Reproduced with permission from Ref. [14]

mentioning that water tetramers can form spontaneously by heating up the sample to 50–80 K, at which the water monomers can gain enough thermal energy to overcome the diffusion barrier. Such tetramers have exactly the same structure and adsorption configuration as the ones built by manipulation at 5 K.

In the following, we show the possibility of applying the orbital-imaging technique to water tetramers. The water tetramer appears as a featureless square protrusion at a large tip-molecule separation (Fig. 3.9a), which splits into four equivalent lobes as the tip height decreases (Fig. 3.9b). The geometric center of the tetramer is right above the Cl^- with the four water molecules adsorbed on the Na^+ (Fig. 3.9b). The calculated most stable tetramer structure (Fig. 3.9c, d) shows that each water molecule donates and accepts just one H-bond yielding a cyclic tetramer. The other four free OH bonds point obliquely upward away from the surface. The observed flat tetramer is more stable than the buckled one predicted by Yang et al. [29].

Such a cyclic water tetramer contains two degenerated chiral states, clockwise and anticlockwise H-bonding loops. In a previous work, the chirality of H-bonded methanol hexamer was indirectly deduced by combining STM with DFT simulations [35]. However, it has been not possible to directly visualize the O–H directionality and the associated chirality of water nanoclusters. Based on the mechanism of orbital imaging technique, we decrease the tip height to increase the tip-water coupling and found that the four lobes of the tetramer become no longer equivalent but slightly dis-

3.3 Submolecular-Resolution Imaging of Interfacial Water with STM 55

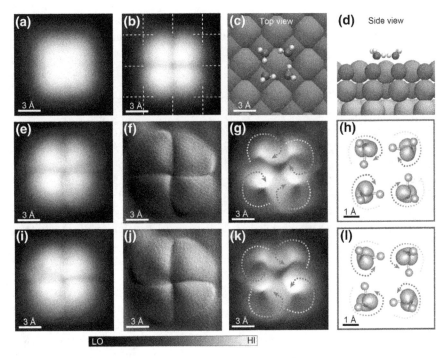

Fig. 3.9 Orbital imaging of water tetramers. a, b STM images of a water tetramer acquired at different tunneling gaps. The white square grid in (**b**) denotes the sub-lattice of Cl^-. **c, d** Calculated adsorption structure of a tetramer. **e–l** HOMO imaging of a water tetramer with two different chiral states, anticlockwise (**e–h**) and clockwise (**i–l**). H-bonded loops with different tip-water coupling strength. **h, l** Calculated HOMO of the two chiral tetramers by plotting isosurfaces of charge densities integrated over 1 eV of the HOMO tail close to E_F. Set point of the STM images: **a** 40 mV, 10 pA; **b** 10 mV, 50 pA; **e** and **i** 10 mV, 80 pA; **g** and **k** 10 mV, 140 pA. The STM images were obtained at 5 K. Reproduced with permission from Ref. [14]

torted, such that the boundaries between the four lobes exhibit left-handed (Fig. 3.9e) or right-handed (Fig. 3.9i) rotation. The chirality of the tetramer appears more evident in the corresponding derivative images (Fig. 3.9f, j).

Further increasing the tip-water coupling, surprisingly, each lobe of the tetramer appears as helical structures highlighted by curved dotted arrows (Fig. 3.9g, k). Such helical structures resemble the HOMO of water molecule, which shows the same chiral character because of the tilt of the HOH plane from the surface normal (Fig. 3.9h, l). Indeed, the HOMO imaging of water tetramer is further confirmed by the calculated PDOS of water tetramer adsorbed on NaCl(001), which suggests the broadening and shift of the HOMO towards E_F as a result of tip-water coupling, whereas the LUMO is less effected and stays above 1 eV [14]. Comparison between the orbital images and the HOMO isosurfaces allows us to identify the directionality of H-bonds and discriminate the H-bonding associated chirality of water tetramer, which provides further opportunities for probing the dynamics of H-bonded networks at the atomic scale such as H-atom transfer and bond rearrangement.

3.3.4 Characterization of H-Bonded Water Nanoclusters

To obtain bigger water clusters on the Au-supported NaCl(001) surface, we dose water molecules with the temperature of substrate above 50 K. Figure 3.10a shows a typical STM image focusing on a NaCl island after dosing about 0.1 bilayer water molecules at 77 K. Due to the high mobility of water on the NaCl(001) surface, most of the water molecules aggregate at the edge of the NaCl island while isolated water clusters can form on the terrace occasionally. We found four typical types of water clusters with regular shape, denoted as I, II, III and IV in Fig. 3.10. In addition, larger water clusters with irregular shape were also observed. In order to identify the structure of these nanoclusters, we performed high-resolution STM imaging and ab initio DFT calculations.

Figure 3.11 shows the high-resolution STM images and the calculated adsorption configurations of the four types of water clusters. The type-I (Fig. 3.11a) water cluster corresponds to a water tetramer, which is the same as the one constructed at 5 K. The water tetramer is the most frequently discovered cluster on the surface and acts

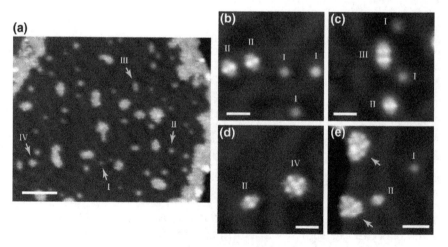

Fig. 3.10 Growth of water nanoclusters on NaCl(001) islands. a STM topography of a NaCl(001) island after dosing about 0.1 bilayer water molecules at 77 K. The water clusters appear in four characteristic representations (denoted as I, II, III, and IV by the sky-blue arrows). Set point: V = 500 mV and I = 10 pA. **b–e,** Zoom-in STM images of the isolated water clusters I, II, III, and IV. The larger water clusters with irregular shapes are denoted by the sky-blue arrows in (**e**). **b** and **d** were acquired at 77 K. **c** and **e** were acquired at 5 K. Set points of **b**: V = 20 mV and I = 10 pA, **c**: V = 100 mV and I = 10 pA, **d**: V = 20 mV and I = 150 pA, **e**: V = 100 mV and I = 10 pA. Reproduced with permission from Ref. [16]

3.3 Submolecular-Resolution Imaging of Interfacial Water with STM 57

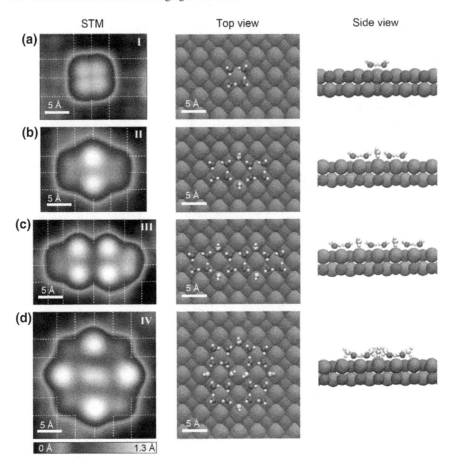

Fig. 3.11 High-resolution STM images and schematic models of four types of water clusters. **a** Type-I; **b** type-II; **c** type-III; and **d** type-IV. The white square grid overlaid on the STM images denotes the sub-lattice of Cl^-. H, Cl^- and Na^+ are denoted by white, grey and dark-cyan spheres, respectively, in the adsorption configurations of the water clusters. The O atoms of water molecules in upper layers are represented by yellow spheres and the lower layers are red spheres. Set point **a** 20 mV, 50 pA; **b** 7 mV, 550 pA; **c** 6 mV, 400 pA; **d** 5 mV, 120 pA. Reproduced from with permission Ref. [16]

as the basic building block to form larger water clusters. The STM images of type-II (Fig. 3.11b), -III (Fig. 3.11c) and -IV (Fig. 3.11d) water clusters exhibit interesting bilayer structures, in which the tetramers in the lower layers are connected by bridging water molecules in the upper layers. Such a bridging mechanism was confirmed by the DFT calculations. The type-IV water cluster consists of four tetramers, which are interconnected by six bridging water molecules, four at the periphery and two in the center (Fig. 3.11d). Interestingly, the pair of inner bridging water molecules results in the formation of a Bjerrum D-type defect [36, 37], where there are two protons between the nearest neighbor oxygen atoms, but they avoid facing each other due to the repulsive interactions between the two upward H atoms. Such Bjerrum D-type defects have also been observed in water-hydroxyl layers on Cu(110) [37, 38]. The other OH bond of each bridging water molecule points downward, forming H bond with the Cl^- anion so that the Cl^- is lift ~0.5 Å from the NaCl surface. The lifting of the Cl^- via the H^+–Cl^- interaction suggests that the Cl^- tends to depart prior to the Na^+ at the early stage of the salt dissolution, which is consistent with the earlier theoretical predictions [39–41].

3.3.5 An Unconventional Bilayer Ice

With increasing water coverage, water molecules will eventually wet the NaCl(001) surface and form large-scale 2D ice islands via the bridging mechanism (Fig. 3.12a). The magnified STM image of 2D ice appears as an array of paired protrusions with two orthogonal orientations, one of which is denoted by a dashed white ellipse (Fig. 3.12b). As an extension of type-IV cluster, the 2D ice overlayer shows tetragonal bilayer structure, where the tetramer arrays in the lower layer are interconnected by paired bridging water molecules within the upper layer, giving rise to a large density of Bjerrum-D type defects (Fig. 3.12c). Those Bjerrum-D type defects provide a large number of dangling OH bonds, facilitating the nucleation of second-layer ice highlighted by cyan arrows in Fig. 3.12a.

Such a tetragonal bilayer ice structure built from cyclic water tetramers goes well beyond the conventionally simple hexagonal bilayer model. Recently, similar square ice structure was also observed between the layers of graphene bilayer due to the hydrophobic confinement [42]. Notably, the formation of the periodic Bjerrum-D type of defects with unusually high density in the 2D ice strongly violates the Bernal-Fowler-Pauling ice rules [43] and may play a crucial role in catalyzing heterogeneous chemical reactions on water-coated salt surfaces as well as in influencing various phenomena such as heterogeneous ice nucleation, salt dissolution and caking.

3.4 Non-invasive Imaging of Interfacial Water with AFM

3.4.1 Background

Except for the extensive studies of water-solid interfaces with STM, AFM has also been an ideal tool to visualize the microscopic structure and dynamics of water at solid surfaces. Water adlayers on mica coated by graphene at ambient environment are observed by AFM, where the first water layer shows ice-like phase with the height of 0.37 nm, resembling the height of the ice monolayer [44]. With the development of three dimensional AFM (3D-AFM), the vertical and lateral distribution of water hydration are resolved with atomic resolution [45]. More recently, 3D-AFM enables

3.4 Non-invasive Imaging of Interfacial Water with AFM

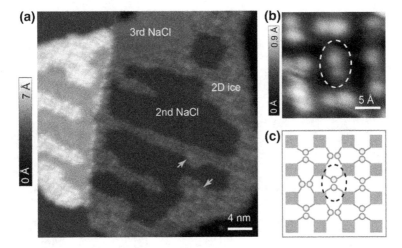

Fig. 3.12 The formation of 2D tetragonal bilayer ice on the NaCl(001) surface. a STM image of the 2D ice on a NaCl island. Additional water molecules adsorbed on the 2D ice islands, highlighted by the cyan arrows. **b** Zoom-in STM image of 2D ice. **c** Schematic model of the 2D tetragonal bilayer ice, corresponding to (**b**). The red solid squares denote the water tetramers and the yellow spheres denote the bridging water molecules. The dashed white and black ellipses in (**b** and **c**) highlight one of the Bjerrum D-type defect. Set point: **a** 400 mV, 10 pA; **b** 350 mV, 10 pA. The STM images were acquired at 77 K. Reproduced with permission from Ref. [16]

direct visualization of point defects in the hydration structure of calcite (10.4) and quantitative characterization of the influence of the point defect on the surrounding hydration structure [46]. Interfacial water at UHV environment are also investigated with AFM, revealing the growth mode of hexagonal and cubic ice at low temperature [47]. Moreover, ultra-high resolution imaging of water networks on Cu(110) surface have been achieved with q-Plus based nc-AFM [38].

However, an intrinsic problem of SPM is that all the probes inevitably induce perturbation to the fragile water structure, due to the excitation of the tunneling electrons and the tip-water interaction forces. The atomic resolution of organic molecules is typically achieved at the very small tip-molecule separation where the short-range Pauli repulsion force is dominant [9, 24, 48, 49]. The tip-molecule interaction in this range is quite strong such that significant relaxation of the tip apex is induced [24]. Considering that H bonds are much weaker than covalent bonds, the water structure may be easily disturbed at small tip heights [38]. At large tip heights where only the long-range van der Waals and electrostatic forces are detectable, the resolution is usually quite poor for weakly polarized molecules. However, in contrary to the weakly polarized aromatic molecules, the water molecule has a strong internal dipole moment. Therefore, the imaging mechanism driven by the electrostatic force greatly relies on the detailed charge nature of the tip apex [50, 51].

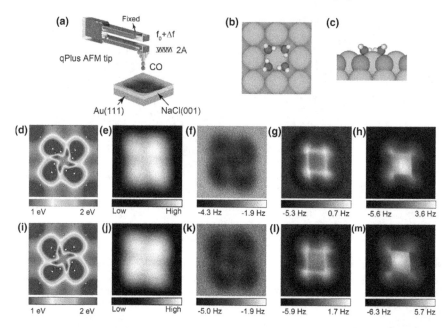

Fig. 3.13 Experimental set-up and AFM images of two degenerate water tetramers with a CO-terminated tip. a Schematic of a qPlus-based nc-AFM with a CO-tip. b, c Top and side view of the water tetramer adsorbed on the NaCl(001) surface, respectively. H, O, Cl, Na atoms are denoted as white, red, green and purple spheres, respectively. d–h and i–m Water tetramers with clockwise and anticlockwise H-bonded loops, respectively. d, i Calculated electrostatic potential map of the water tetramers. e, j Constant-current STM images with CO-tip. f–m Experimental Δf images recorded at the tip heights of 100 pm, 10 pm, −40 pm, respectively. The tip height is referenced to the STM set point on the NaCl surface (100 mV, 50 pA). The oscillation amplitude is 100 pm. The size of the images is 1.2 nm × 1.2 nm. Reproduced with permission from Ref. [15]

3.4.2 AFM Images of Two Degenerate Water Tetramers with a CO-Terminated Tip

The experimental set-up is schematically shown in Fig. 3.13a, where the tip apex is functionalized with a CO molecule (See Methods in Sect. 3.2). Water tetramers were constructed by assembling four individual H_2O monomers on the NaCl(001) surface at 5 K (Fig. 3.8). The STM images (Fig. 3.9) reveal that each water molecule acts as a single H-bond donor and single H-bond acceptor resulting in a cyclic tetramer (Fig. 3.13b), whereas the other four free OH bonds point obliquely upward away from the surface (Fig. 3.13c) [14]. In fact, the cyclic water tetramer contains two degenerate chiral H-bonded loops, which are respectively displayed in Fig. 3.13d, i, with the calculated Hartree potential superimposed. With a CO-terminated tip, the H-bonding chiralities could not be distinguished from the corresponding STM images (Fig. 3.13e, j).

3.4 Non-invasive Imaging of Interfacial Water with AFM

Then we obtained the constant-height Δf images with the CO-tip at three different tip heights (Fig. 3.13f–h, k–m). At a large tip height, the two tetramers were imaged as four "ear-like" depressions with distinct chirality, which closely resemble their electrostatic potential (Fig. 3.13d, i). As the tip height decreased, the H-bonded loop was visualized as a bright square (Fig. 3.13g, l), which denotes the oxygen skeletons. When further approaching the tip, besides the sharpening of the square lines, contrast inversion was also observed at the center of the tetramer (Fig. 3.13h, m). However, it is very interesting and unusual that the chiral submolecular contrast only appears at large tip heights and almost vanishes at small tip heights.

3.4.3 The Role of Electrostatics in the High-Resolution AFM Imaging of Water Tetramers

To explore the imaging mechanism of submolecular contrast obtained at large tip height, where the long-range force dominates the tip-water interaction, we simulate the AFM images with a molecular mechanics model including the electrostatic force, based on the methods described in refs. [24] and [25]. Figure 3.14a–d display the simulated AFM images of a water tetramer with neutral, s, p_z, d_{z^2} tip models, respectively, at different tip heights z_1, z_2, and z_3 as denoted in Fig. 3.14e. It is obvious that the simulated AFM Δf images with the neutral tip model (Fig. 3.14a, z_2 and z_3) agree well with the experimental results at small tip heights (Fig. 3.13l, m). The sharpening and inversion behavior at small tip height is consistent with the AFM imaging properties of aromatic molecules, resulting from the Pauli repulsion and the consequent lateral relaxation of the CO molecule at the tip apex [9, 52].

However, AFM images with neutral tip at large tip height exhibit square depression feature, which fails to reproduce the experimental AFM feature. Then, we focus the simulated AFM images with s, p_z, d_{z^2} tip models (Fig. 3.14b–d). The simulated images with the monopole and dipole tips at the large tip height (Fig. 3.14b and c, z_1) show very little chirality. Surprisingly, the simulation results with the quadrupole tip (Fig. 3.14d, z_1) perfectly reproduce the AFM images, including the "ear-like" chiral features (Fig. 3.13k) can, as well as the sharp lines and contrast inversion feature at the small tip heights (Fig. 3.14d, z_2 and z_3). The quadrupole nature of the CO-tip can be verified from charge redistribution of the adsorbed CO on metal tip [51] (Fig. 3.14f).

As a matter of fact, the variation of the AFM contrast using different tip models can be understood from the analysis of calculated electrostatic forces acting between the sample and the given tip model. As shown in Fig. 3.14g, the xz-cut planes of vertical electrostatic force present significantly different shapes and decay behaviors for different charged tip models. To extract the contribution of electrostatic force more clearly, we plotted the calculated force curves with s, p_z and d_{z^2} tips after subtraction of the force with a neutral tip (Fig. 3.15). Approximatively, we used an exponential fitting to obtain the decay length of the electrostatic force between the

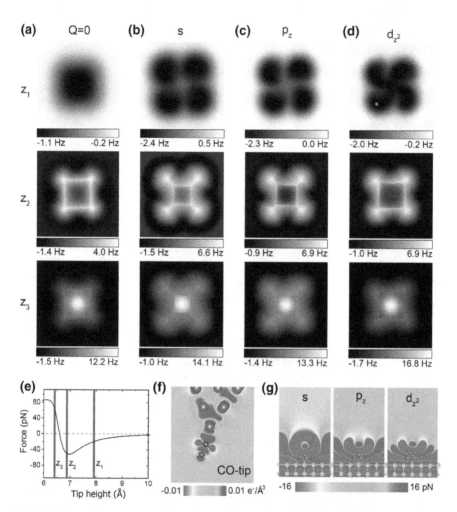

Fig. 3.14 The role of electrostatics in the high-resolution AFM imaging of a water tetramer.
a–d Simulated AFM images of a water tetramer with neutral, s, p_z and d_{z^2} tip models (k = 0.5 N/m, Q = −0.2 e) at three different tip heights. The tip height is defined as the distance between the outmost metal atom of the tip and the upward H atom of the water tetramer. The oscillation amplitude is 100 pm. The size of the images is 1.2 nm × 1.2 nm. **e** Simulated force curve of the water tetramer taken with the d_{z^2} tip, where the three tip heights (z_1, z_2 and z_3) are denoted. **f** Charge distribution of the CO-tip from DFT calculations. **g** Maps of calculated vertical electrostatic forces between the sample and different tip models (s, p_z and d_{z^2}) computed by convolution of Hartree potential of sample and model charge distribution on the tip. Reproduced with permission from Ref. [15]

3.4 Non-invasive Imaging of Interfacial Water with AFM

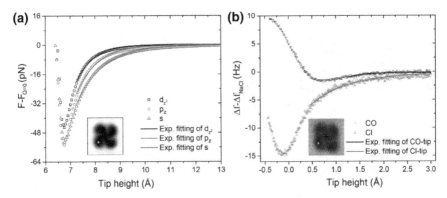

Fig. 3.15 The force curves between the water tetramer and different tips. a The calculated force curves with s, p_z and d_{z^2} tips after subtraction of the force with a neutral tip. The tip position is indicated with a star in the inset. The solid lines are the corresponding exponential fittings of the curves within the range of tip height from 7.2 Å to 13 Å. **b** The frequency shift Δf measured above the water tetramer (as indicated with a star in the inset) with CO- and Cl-tips after removing the contribution from the NaCl substrate. The solid lines are the corresponding exponential fittings of the curves within the range of tip height from 0.45 Å (Cl-tip) or 1.25 Å (CO-tip) to 3 Å. The decay lengths of different tips are summarized in Table 3.1. Reproduced with permission from Ref. [15]

Table 3.1 The fitted decay length of force curves obtained with different tips

Calculated decay length (Å)			Experimental decay length (Å)	
s tip	p_z tip	d_{z^2} tip	Cl-tip	CO-tip
1.039 ± 0.006	0.814 ± 0.005	0.654 ± 0.004	0.786 ± 0.014	0.326 ± 0.03

Reproduced with permission from Ref. [15]

tetramer and different tips. To avoid the effect of tip relaxation at short tip-water separation, only the data points at large tip heights were fitted. The decay length of the d_{z^2} tip is the smallest, as shown in Table 3.1. Similarly, we exponentially fitted the experimental Δf curves with CO- and Cl-tips after removing the contribution from the NaCl substrate (Fig. 3.15b). We found that the decay length with the Cl-tip is more than two times larger than that with the CO-tip (Table 3.1), indicating the short-range nature of the high-order electrostatic force between the CO-tip and the water molecules.

In order to verify the proposed imaging mechanism above, we functionalized the tip apex with a Cl atom, which is negatively charged with about 0.3–0.4 e [17, 53] when attached to the metal tip, acting as a monopole tip. Figure 3.16b, g are the constant-height Δf images at a large tip height, showing negligible chirality. This is consistent with the AFM simulation using the monopole tip (Fig. 3.14b, z_1), revealing the low sensitivity of monopole-like probe charges for high-resolution mapping of complex electrostatic fields. At smaller tip heights, the Δf images

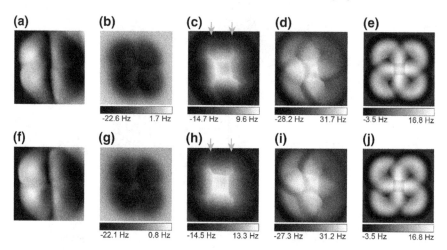

Fig. 3.16 Experimental and simulated AFM images of two degenerate water tetramers with a Cl-terminated tip. **a, f** Derivative STM images of the water tetramers with clockwise and anticlockwise hydrogen-bonded loops, respectively. Set point: 10 mV and 50 pA. **b–d** and **g–i** Corresponding Δf images. All the images were obtained on the same tetramer (switched) with the same Cl-tip. The tip heights are 30 pm (**b** and **g**), −120 pm (**c** and **h**), −120 pm (**d** and **i**). The oscillation amplitudes are 40 pm (**b** and **g**), 100 pm (**c** and **h**), 40 pm (**d** and **i**). The "fork-like" features are denoted by two green arrows in (**c**) and (**h**). **e, j** Simulated Δf images with the oscillation amplitudes of 40 pm, which were obtained with a monopole (s) tip (k = 0.5 N/m, Q = −0.25 e). The size of the images is 1.2 nm × 1.2 nm. Reproduced with permission from Ref. [15]

(Fig. 3.14c, h) show prominent sharp squares and "fork-like" features at the periphery (see the green arrows), also agreeing well with the simulation (Fig. 3.14b, z_2). When simulating with a smaller oscillation amplitude, the Δf images change remarkably, showing bright helical structures with distinct chirality (Fig. 3.16d, e and i, j). Such helical features arise from the pronounced tip relaxation at close tip-water distances, which is determined by the complex interplay between the Pauli and the electrostatic interaction [54].

In the case of large oscillation amplitude, the probe spends large part of the oscillation period at tip-sample distances, where the chirality of the electrostatic potential is almost negligible. In addition, the electrostatic potential changes significantly at the very close distance, having a non-trivial 3D chiral character. In the limit of the small amplitude, the frequency shift is proportional to derivative of force along z-distance. Therefore, the non-trivial 3D character of the electrostatic potential induces a significant impact on the frequency shift when small amplitude is used.

3.4.4 Submolecular-Resolution AFM Images of Weakly Bonded Water Clusters

Comparing with the short-range Pauli repulsion interaction, the long-range high order electrostatic interaction between the quadrupole CO tip and the dipole water molecule is very small such that the disturbance of the CO-tip on the water structure should be minimal, which may open up the possibility of probing weakly bonded water clusters other than the rigid tetramers based on the above described submolecular-resolution AFM imaging at large tip height. To confirm this possibility, we investigated fragile water structures such as dimers and trimers, which are not stable and very difficult to image with STM.

Figure 3.17a–d are the geometric structures, experimental STM images, experimental and simulated Δf images of three water dimers at large tip heights, respectively. Similarly, we found that the depression features directly reflect the distribution of electrostatic potential in the water dimers. It is worthy to be noted that the crooked depressions in the AFM images are actually correlated with the position of the H atoms, which can help us identify the detailed configuration of various water clusters with unprecedented precision. It is striking that the AFM imaging can readily distinguish the subtle difference of the O-H tilting in the water dimers, which is inaccessible with the STM images (Fig. 3.17b). Water trimers are even more unstable than the dimers since they can have many metastable states and the energy variations are smaller than 50 meV, but we are still able to image the electrostatic potential of various water trimers with submolecular resolution (Fig. 3.17e–h). In combination with the simulations, their atomic configurations can be unambiguously determined.

This technique is also applicable for more complicated water structures such as bilayer triple-tetramers (Fig. 3.18), which is composed of three tetramers which are bridged with four standing water molecules, forming a bilayer ice cluster (Fig. 3.18a, b) [16]. The bridging water molecules were imaged as four bright spots at the large tip height (Fig. 3.18c), which result from the Pauli repulsion force between the CO-tip and the standing water molecules. The chirality of the tetramers at two ends within the bottom layer of the bilayer ice can be clearly resolved, which has been not possible with STM before [16]. Surprisingly, the chirality of the middle tetramer can be also distinguished although it is somewhat blocked by the higher bridging water molecules. The skeleton of the H-bonding network in the triple-tetramer can be seen very clearly under close imaging condition (Fig. 3.18d). The sharp lines emerge from the deflection of the probe particle due to its repulsive interaction with the nearest neighboring water molecules. Similar results were obtained for water overlayers on Cu surfaces recently [38]. The simulated AFM images at large (Fig. 3.18e) and small (Fig. 3.18f) tip heights agree well with the experimental results. Note that the imaging of H-bonding skeleton requires relatively strong tip-water interaction at short range, which may induce significant disturbance to the water structure.

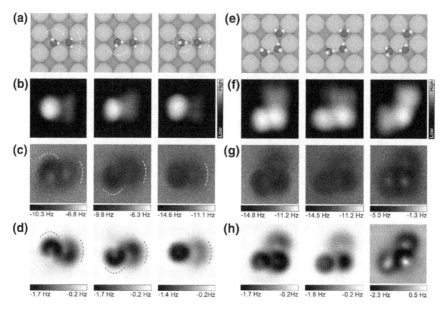

Fig. 3.17 Submolecular-resolution AFM images of weakly bonded water dimers and trimers with a CO-tip. a–d and e–h Geometric structures, experimental STM images, experimental and simulated Δf images of water dimers (a–d) and trimers (e–h), respectively. The crooked depressions in the AFM images are highlighted by dashed lines in a, c and d. Set point of (b): 100 mV and 50 pA, 30 pA, 25 pA (from left to right), respectively. The tip height of (c) is 100 pm, 100 pm and 130 pm (from left to right), respectively. Set point of (f): 100 mV and 25 pA, 20 pA, 15 pA (from left to right), respectively. The tip height of (g) is 130 pm, 130 pm and 110 pm (from left to right), respectively. All the oscillation amplitudes of experimental and simulated images are 100 pm. All the simulations were done with a quadrupole (d_{z^2}) tip (k = 0.5 N/m, Q = -0.2 e). The size of the images: 1.2 nm × 1.2 nm. Reproduced with permission from Ref. [15]

3.4.5 Quantitative Characterization of the Non-invasive AFM Imaging Technique

In order to be more quantitative on how non-invasive of AFM imaging can be, we performed systematic AFM measurements to estimate the minimum forces needed to achieve the submolecular resolution (Fig. 3.19), following the method described in Ref. [55]. We can get F_z and the interaction energy U from the frequency shift Δf at different tip heights according to Ref. [56]. The lateral force F_x (F_y) can be then obtained from the partial differentiation of the interaction energy U along x (y) direction. It can be seen that the minimum vertical (z) force which yields the submolecular resolution of the water dimer is only about 70.9 pN. The minimum lateral force along x (y) direction is 7.9 (10.5) pN. The corresponding tip-water interaction energy is only 43.8 (57) meV, which allows the imaging of metastable water structures with very small transition barrier.

3.4 Non-invasive Imaging of Interfacial Water with AFM

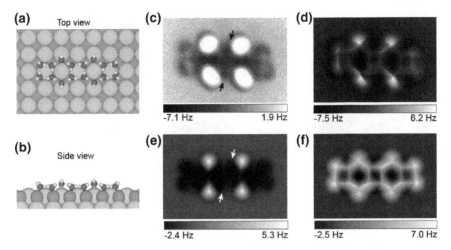

Fig. 3.18 High-resolution AFM images of a water triple-tetramer with a CO-tip. a, b Top and side views of the atomic structures of the water triple-tetramer. **c, d** and **e, f** Experimental and simulated Δf images of the water triple-tetramer, respectively. The tip heights: **c** 110 pm, **d** 20 pm, **e** 6.19 Å, **f** 5.16 Å. The chirality of the central tetramer is denoted by arrows in (**c**) and (**e**). All the oscillation amplitudes of experimental and simulation images are 100 pm. The simulations were done with a quadrupole (d_{z^2}) tip (k = 0.5 N m^{-1}, Q = -0.2 e). The size of the images: 2 nm × 3 nm. Reproduced with permission from Ref. [15]

It is worthy to recall that the Cl-tip can also obtain submolecular-resolution imaging of the electrostatic potential of water tetramer by using small oscillation amplitudes (Fig. 3.16d, i). However, such a resolution is only achieved at small tip-water separation where the electrostatic and Pauli force becomes strong enough to induce significant relaxation of the tip apex. Any attempts to enter into this region can easily disturb the weakly bonded water clusters such as the water dimers, trimers and bilayer ice clusters (Fig. 3.20). Therefore, the high-order electrostatic force between the CO-tip and the water is critical since it yields submolecular resolution at relatively large tip-water separations where the electrostatic force and other force components are still rather weak, thus avoiding the disturbance of the tip on the water molecules.

3.5 Summary

In this chapter, we demonstrate unprecedented submolecular imaging of water on a Au-supported insulating NaCl(001) film with STM and nc-AFM technique. The frontier molecular orbitals of adsorbed water were directly visualized by enhancement of the DOS near E_F via tunable tip-water coupling, which allowed discriminating the orientation of the monomers and the H-bond directionality of the tetramers in real space. Based on submolecular orbital imaging technique, the nucleation and growth of water nanoclusters and overlayer ice are

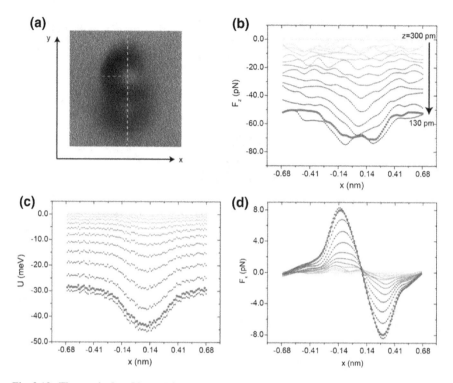

Fig. 3.19 The vertical and lateral force needed to obtain submolecular resolution of a water dimer. a Dashed red (blue) line shows the x (y) direction along which the frequency shift data at different tip heights (z = 300 pm to 130 pm) were obtained. Note that the frequency shifts are weighted time-averaged over the cantilever oscillation between $z = z'$ and $z = z' + 2A$. The amplitude A is 100 pm. **b–d** Vertical force F_z, tip-sample interaction energy U, and lateral force F_x extracted from the frequency shift along x direction. The tip height changes every 5 pm (only some of them are shown here). The red curves in (**b–d**) correspond to the tip height where the submolecular resolution can be obtained. We also carried out similar measurements along the dashed blue line (y direction) as shown in (**a**). The corresponding lateral force and interaction energy needed to obtain submolecular resolution are 10.5 pN and 56 meV, respectively. Reproduced with permission from Ref. [15]

imaged at the molecular level. What's more, the non-perturbative submolecular-resolution AFM images of water obtained by CO-tip not only provide the spatial information of electrostatics, but also allow us to determine the detailed H-bonding structure including the position of the H atoms, which is crucial for the understanding and investigation of H-bonding interaction and dynamics of water.

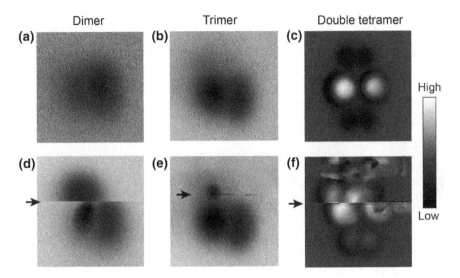

Fig. 3.20 The disturbance of Cl-tip on the water dimer, trimer and double-tetramer. **a, d** Δf images of a water dimer at tip heights of 140 and 120 pm, respectively. **b, e** Δf images of a water trimer at tip heights of 0 and −30 pm, respectively. **c, f** Δf images of a water double-tetramer at tip heights of −50 pm and −100 pm, respectively. The oscillation amplitudes: **a, b, d** and **e** 100 pm; **c, f** 50 pm. The size of the images: **a, b, d** and **e** 1.4 nm × 1.4 nm; **c** and **f** 2 nm × 2 nm. Reproduced with permission from Ref. [15]

References

1. Thiel PA, Madey TE (1987) The interaction of water with solid surfaces: fundamental aspects. Surf Sci Rep 7:211–385
2. Henderson MA (2002) The interaction of water with solid surfaces: fundamental aspects revisited. Surf Sci Rep 46:1–308
3. Verdaguer A, Sacha GM, Bluhm H, Salmeron M (2006) Molecular structure of water at interfaces: Wetting at the nanometer scale. Chem Rev 106:1478–1510
4. Hodgson A, Haq S (2009) Water adsorption and the wetting of metal surfaces. Surf Sci Rep 64:381–451
5. Zou Z, Ye J, Sayama K, Arakawa H (2001) Direct splitting of water under visible light irradiation with an oxide semiconductor photocatalyst. Nature 414:625–627
6. Akiya N, Savage PE (2002) Roles of water for chemical reactions in high-temperature water. Chem Rev 102:2725–2750
7. Kudo A, Miseki Y (2009) Heterogeneous photocatalyst materials for water splitting. Chem Soc Rev 38:253–278
8. Eisenberg DS, Kauzmann W (1969) The structure and properties of water. Clarendon P., Oxford
9. Gross L, Mohn F, Moll N, Liljeroth P, Meyer G (2009) The chemical structure of a molecule resolved by atomic force microscopy. Science 325:1110–1114
10. Albrecht F, Neu M, Quest C, Swart I, Repp J (2013) Formation and characterization of a molecule-metal-molecule bridge in real space. J Am Chem Soc 135:9200–9203
11. Zhang J et al (2013) Real-space identification of intermolecular bonding with atomic force microscopy. Science 342:611–614
12. Kawai S et al (2016) Van der Waals interactions and the limits of isolated atom models at interfaces. Nat Commun 7:11559

13. Hämäläinen SK et al (2014) Intermolecular contrast in atomic force microscopy images without intermolecular bonds. Phys Rev Lett 113:186102
14. Guo J et al (2014) Real-space imaging of interfacial water with submolecular resolution. Nat Mater 13:184–189
15. Peng JB et al (2018) Weakly perturbative imaging of interfacial water with submolecular resolution by atomic force microscopy. Nat Commun 9:112
16. Chen J et al (2014) An unconventional bilayer ice structure on a NaCl(001) film. Nat Commun 5:4056
17. Meng X et al (2015) Direct visualization of concerted proton tunnelling in a water nanocluster. Nat Phys 11:235–239
18. Horcas I et al (2007) WSXM: a software for scanning probe microscopy and a tool for nanotechnology. Rev Sci Instrum 78:013705
19. Kresse G, Hafner J (1993) Ab initio molecular dynamics for liquid metals. Phys Rev B 47:558–561
20. Kresse G, Furthmuller J (1996) Efficient iterative schemes for ab initio total-energy calculations using a plane-wave basis set. Phys Rev B 54:11169–11186
21. Klimes J, Bowler DR, Michaelides A (2011) Van der Waals density functionals applied to solids. Phys Rev B 83:195131
22. Kresse G, Joubert D (1999) From ultrasoft pseudopotentials to the projector augmented-wave method. Phys Rev B 59:1758–1775
23. Henkelman G, Uberuaga BP, Jonsson H (2000) A climbing image nudged elastic band method for finding saddle points and minimum energy paths. J Chem Phys 113:9901–9904
24. Hapala P et al (2014) Mechanism of high-resolution STM/AFM imaging with functionalized tips. Phys Rev B 90:085421
25. Hapala P, Temirov R, Tautz FS, Jelinek P (2014) Origin of high-resolution IETS-STM images of organic molecules with functionalized tips. Phys Rev Lett 113:226101
26. Hebenstreit W et al (1999) Atomic resolution by STM on ultra-thin films of alkali halides: experiment and local density calculations. Surf Sci 424:L321–L328
27. Michaelides A, Ranea VA, de Andres PL, King DA (2003) General model for water monomer adsorption on close-packed transition and noble metal surfaces. Phys Rev Lett 90:216102
28. Cabrera-Sanfelix P, Arnau A, Darling GR, Sanchez-Portal D (2006) Water adsorption and diffusion on NaCl(100). J Phys Chem B 110:24559–24564
29. Yang Y, Meng S, Wang EG (2006) Water adsorption on a NaCl (001) surface: a density functional theory study. Phys Rev B 74:245409
30. Repp J, Meyer G, Stojkovic SM, Gourdon A, Joachim C (2005) Molecules on insulating films: scanning-tunneling microscopy imaging of individual molecular orbitals. Phys Rev Lett 94:026803
31. Ho W (2002) Single-molecule chemistry. J Chem Phys 117:11033–11061
32. Kumagai T (2015) Direct observation and control of hydrogen-bond dynamics using low-temperature scanning tunneling microscopy. Prog Surf Sci 90:239–291
33. Norskov JK (1990) Chemisorption on metal surfaces. Rep Prog Phys 53:1253–1295
34. Martinez JI, Abad E, Gonzalez C, Flores F, Ortega J (2012) Improvement of scanning tunneling microscopy resolution with H-sensitized tips. Phys Rev Lett 108:246102
35. Lawton TJ, Carrasco J, Baber AE, Michaelides A, Sykes ECH (2011) Visualization of hydrogen bonding and associated chirality in methanol hexamers. Phys Rev Lett 107:256101
36. Bjerrum N (1952) Structure and properties of ice. Science 115:385–390
37. Forster M, Raval R, Hodgson A, Carrasco J, Michaelides A (2011) c(2 × 2) Water-hydroxyl layer on Cu(110): a wetting layer stabilized by Bjerrum defects. Phys Rev Lett 106:046103
38. Shiotari A, Sugimoto Y (2017) Ultrahigh-resolution imaging of water networks by atomic force microscopy. Nat Commun 8:14313
39. Cabrera-Sanfelix P et al (2007) Spontaneous emergence of Cl^- anions from NaCl(100) at low relative humidity. J Phys Chem C 111:8000–8004
40. Klimes J, Bowler DR, Michaelides A (2013) Understanding the role of ions and water molecules in the NaCl dissolution process. J Chem Phys 139:234702

References

41. Liu L-M, Laio A, Michaelides A (2011) Initial stages of salt crystal dissolution determined with ab initio molecular dynamics. Phys Chem Chem Phys 13:13162–13166
42. Algara-Siller G et al (2015) Square ice in graphene nanocapillaries. Nature 519:443–445
43. Bernal JD, Fowler RH (1933) A theory of water and ionic solution, with particular reference to hydrogen and hydroxyl ions. J Chem Phys 1:515–548
44. Xu K, Cao P, Heath JR (2010) Graphene visualizes the first water adlayers on mica at ambient conditions. Science 329:1188–1191
45. Fukuma T, Ueda Y, Yoshioka S, Asakawa H (2010) Atomic-scale distribution of water molecules at the mica-water interface visualized by three-dimensional scanning force microscopy. Phys Rev Lett 104:016101
46. Songen H et al (2018) Resolving point defects in the hydration structure of calcite (10.4) with three-dimensional atomic force microscopy. Phys Rev Lett 120:116101
47. Thurmer K, Nie S (2013) Formation of hexagonal and cubic ice during low-temperature growth. Proc Natl Acad Sci U S A 110:11757–11762
48. Gross L et al (2010) Organic structure determination using atomic-resolution scanning probe microscopy. Nat Chem 2:821–825
49. Moll N, Gross L, Mohn F, Curioni A, Meyer G (2010) The mechanisms underlying the enhanced resolution of atomic force microscopy with functionalized tips. New J Phys 12:125020
50. Hapala P et al (2016) Mapping the electrostatic force field of single molecules from high-resolution scanning probe images. Nat Commun 7:11560
51. Ellner M et al (2016) The electric field of CO tips and its relevance for atomic force microscopy. Nano Lett 16:1974–1980
52. Chiang C-l XuC, Han Z, Ho W (2014) Real-space imaging of molecular structure and chemical bonding by single-molecule inelastic tunneling probe. Science 344:885–888
53. Guo J et al (2016) Nuclear quantum effects of hydrogen bonds probed by tip-enhanced inelastic electron tunneling. Science 352:321–325
54. Giessibl FJ (2001) A direct method to calculate tip-sample forces from frequency shifts in frequency-modulation atomic force microscopy. Appl Phys Lett 78:123–125
55. Ternes M, Lutz CP, Hirjibehedin CF, Giessibl FJ, Heinrich AJ (2008) The force needed to move an atom on a surface. Science 319:1066–1069
56. Sader JE, Jarvis SP (2004) Accurate formulas for interaction force and energy in frequency modulation force spectroscopy. Appl Phys Lett 84:1801–1803

Chapter 4
Single Molecule Vibrational Spectroscopy of Interfacial Water

4.1 Background

Vibrational spectroscopy has long been employed to characterize the intramolecular and intermolecular interactions of water [1–8], providing a sensitive probe for nuclear quantum effects (NQEs) of protons in energy space through isotope substitution experiments. Laser-based techniques, such as infrared and Raman spectroscopy have been extensively applied to characterize the H-bonding interactions and NQEs of water in gas and condensed phases [4, 9, 10]. In addition, SFG even enabled the investigation of interfacial water at the molecular level [11]. In spite of the advantage of compatible to various environments, ranging from gas, liquid water to solid ice, these laser-based vibrational spectroscopic techniques suffer from the limitation of poor spatial resolution, averaging effect and spectra assignment, thus prohibiting the accurate and quantitative description of the NQEs.

On the other hand, electron spectroscopy could probe molecular vibrations as well via electron-vibration coupling, for example, electron energy loss spectroscopy (EELS) and inelastic electron tunneling spectroscopy (IETS) (see Chap. 2 for more details). Compared with real-space imaging, vibrational spectroscopy of water with STM is premature and much less used. However, it can offer new insights into the H-bonding configurations, dynamics and strength, especially peering into the degree of freedom of hydrogen. In 1988, Stipe et al. pushed the sensitivity of IETS down to single molecule level using STM [12]. However, most of STM-IETS measurements have been performed on molecules whose frontier orbitals are located far away from the E_F [13–15]. In such off-resonance cases, the electron-vibration coupling is just a weak perturbation on the elastic scattering picture, leading to a very small cross section for the vibration excitation and weak IET signals with the differential conductance change of a few percent [16, 17].

This limitation is particularly true for close-shell molecules like water. Consequently, vibrational spectroscopy of interfacial water by STM has been rarely reported. Morgenstern et al. demonstrated the first IETS of water submonolayer on

Ag(111) and assigned the IET features to various vibrational modes of the water molecule [18]. However, the IET signals were too weak to identify the vibrational modes accurately because of the wide HOMO-LUMO gap of water molecules. Later on, Kumagai et al. acquired IET spectra of the hydroxyl group and its clusters, which enabled precise identification of OH/OD bending and stretching modes [19–21]. The intrinsic problem of the IET spectra is that the prominent vibrational features are mainly attributed to the switching motion between different structures of hydroxyl groups. Hence, it is still a great challenge to obtain reliable and precise vibrational spectroscopy of water molecules with STM-IETS, especially at the single molecule level.

4.2 Resonantly Enhanced IETS

The magnitude of IETS signal can be drastically enhanced when the molecular level lies near the chemical potential of the electrode [22, 23]. In this case, the molecular level directly involves in the tunneling process. Both the tunneling density of states and the e-vib interaction increase, thus the IETS signals are greatly enhanced. In fact, this resonantly enhanced IETS was firstly proposed by Persson and Baratoff theoretically as a possible way to observe IETS in single molecular junction and a differential conductance change of 10% or more could be achieved [22]. Its experimental realization was only made possible in 2009 by Song et al. in a three-terminal single-molecule device by tuning the location of HOMO with respect to the E_F via voltage gating [24]. In a similar way, we show the possibility of gating the frontier orbitals of a water monomer on NaCl(001) surface with a functionalized STM tip through tip-molecule coupling. The signal-to-noise ratio of the tip-enhanced IETS [25] is increased by orders of magnitude over the conventional STM-IETS.

4.3 Methods

The STM measurements were all performed at 5 K. The scanning tunneling spectroscopy, dI/dV and d^2I/dV^2 spectra, were acquired simultaneously using a lock-in amplifier by demodulating the first and second harmonics of the tunneling current, respectively. A modulation voltage of 5–7 mV_{rms} at 237 Hz was added with the feedback loop open.

To understand the IETS signal theoretically, one needs to study the electronic transport problem taking into account electron's interaction with molecular vibrations. The experimental IETS of water monomer is analyzed with the transport calculations based on DFT in combination with nonequilibrium Green's functions developed in Ref. [26]. We calculate the electronic structure using SIESTA [27], transport properties using TranSIESTA [28], and electron-vibration coupling and IETS using Inelastica [29].

4.4 Tip-Enhanced IETS of Water Monomers

4.4.1 Selective Orbital Gating of Water Monomers with Cl-Tip

The STM-IETS experiments were performed on water monomers adsorbed on a Au-supported NaCl(001) bilayer surface. The water monomer adsorbed on the top of Na$^+$ site in a "standing" configuration with one OD dangling upward, the other OD forming a H bond with the Cl$^-$ of the NaCl surface (Fig. 4.1a) [30]. As discussed in Chap. 3, the STM tip not only acts as a probe, but also could modify the molecular DOS around the E_F via tip-molecule electronic coupling. This actually offers a viable way to tune the off-resonance IET process into near-resonance case.

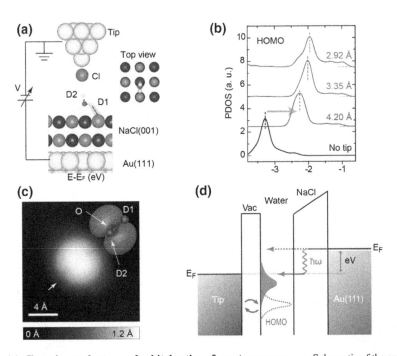

Fig. 4.1 Experimental setup and orbital gating of a water monomer. a Schematic of the experimental setup. Water monomer adsorbed on the Au-supported NaCl(001) surface with a "standing" configuration with a dangling OD (D2) and a H-bonded OD (D1, denoted by dashed line). O, D, Au, Cl$^-$, and Na$^+$ are denoted by red, white, golden, green, and purple spheres, respectively. **b** Calculated PDOS of water monomer with and without STM tip. The position the peaks are highlighted by dashes lines. HOMO moves towards the E_F when the tip appears (gray arrow). **c** STM images of HOMO. Inset is the calculated isosurface of charge density of the HOMO. **d** Schematic of the tip-enhanced IET process. The tip-water coupling tunes the HOMO to the proximity of the E_F, leading to near resonantly-enhanced IETS. Reproduced with permission from Ref. [25]

However, we noticed that the terminations of the bare STM tips are usually uncontrollable, which will influence the gating efficiency of different molecular orbitals [30]. In order to obtain reliable IET spectra in a well-controlled fashion, we use Cl-terminated tip, which is sensitive to HOMO and can selectively enhance the HOMO states around E_F (Fig. 4.1b), while the LUMO is less affected. Compared to the bare metal tip, the orbital gating with a Cl-tip is more efficient because of the strong coupling between the HOMO with the Cl p_z orbital. The highly selectivity of Cl-tip is further confirmed by the corresponding orbital imaging, with which the STM images of water monomers always exhibited a HOMO-like double-lobe structure (Fig. 4.1c), while the LUMO was not observed throughout the accessible bias range (from −400 to 400 mV). In such a near-resonance case (Fig. 4.1d), the electron-vibration interaction is resonantly enhanced, leading to the increase of IET cross section. Considering the key role of tip gating in enhancing the IET signals, we named this technique as tip-enhanced IETS.

4.4.2 Single Molecule Vibrational Spectroscopy of Water Monomers

With the Cl-tip positioned slightly away from the nodal plane of the molecule, we obtained the tip-enhanced IETS of water monomer D_2O (Fig. 4.2). At large tip height, the spectra are featureless (blue curves), simply following the background NaCl signal (grey curves). When approaching the tip by 80 pm toward the water molecule, additional kinks emerge in the dI/dV spectrum (red). These features are more prominent in the corresponding d^2I/dV^2 spectrum and appear as peaks and dips in point symmetry with respect to the zero bias, which are assigned the spectral features to the frustrated rotational ("R", 55 meV), bending ("B", 147 meV), and stretching mode ("S", 326 meV), respectively.

We noticed that the tip-enhanced IETS is very sensitive to the nature of the tip apex (Fig. 4.3). Different tip terminations may selectively gate different molecular orbitals (with different symmetries) to E_F. According to the IETS propensity rules [31, 32], only the vibrational modes whose symmetry match those of the molecular orbitals around E_F are detectable in the tip-enhanced IETS. This actually provides a viable way to activate those vibrational modes which are forbidden in conventional IETS.

4.4.3 Lineshape Change of Tip-Enhanced IETS

To explore the properties of tip-enhanced IETS systematically, we measure the high-resolution IETS of water bending mode at different tip-molecule distances (Fig. 4.4a). With decreasing tip height, the intensity of the normalized IET feature is enhanced

4.4 Tip-Enhanced IETS of Water Monomers

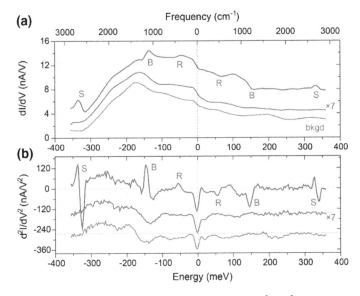

Fig. 4.2 Tip-enhanced IETS of a D$_2$O monomer. a dI/dV. b d^{2}I/dV^2 spectra obtained at tip heights of −40 pm (blue) and −120 pm (red). Gray curve is the background NaCl signal acquired at the tip height of −120 pm. Tip heights are referenced to the gap set with V = 100 mV and I = 50 pA. "R", "B", "S" represent water rotational, bending and stretching vibration mode, respectively. Reproduced with permission from Ref. [25]

by more than one order of magnitude. Meanwhile, the line shape changes gradually from a symmetric dip to an asymmetric Fano-like shape [33]. In order to understand the underlying physical picture of the apparent two features, we simulated the IETS of a D$_2$O monomer at different tip heights using the DFT-based transport theory [26, 28] (Fig. 4.4b), which is in excellent agreement with the experimental data.

Detailed analysis indicates that both inelastic and new high-order elastic channels (Fig. 4.4c–e) become open and determine the lineshape and spectral intensity of IETS signal. The elastic process, involving phonon emission and re-adsorption, can be further divided into two different processes (Fig. 4.4d, e). Now we analyze how the three processes determine the intensity and shape of the IETS signal.

Inelastic process (Fig. 4.4c) opens an additional transport channel, and contributes to a peak in the d^{2}I/dV^2 signal for positive bias. In contrast, the two high-order elastic processes (Fig. 4.4d, e) may contribute to a dip due to their interference with the zero-order elastic tunneling. In the near-resonance case, the high-order elastic processes become more prominent than the inelastic process, resulting in the dip feature. The observed dramatic increase of the spectral intensity arises from the enhanced HOMO DOS with decreasing tip-molecule separation. On the other hand, the elastic processes (Fig. 4.4d, e) also contribute to the asymmetric lineshape. But their contributions are opposite to each other and normally cancel out. Considering that the resonant state (HOMO) of water couples more strongly to the tip, the process

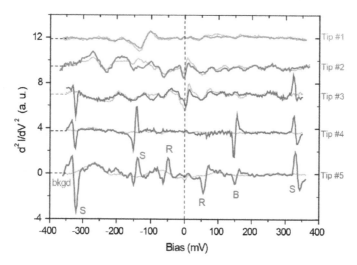

Fig. 4.3 IETS of a D_2O monomer with different types of tips. The tip-enhanced IETS is very sensitive to the nature of tip apex, showing selective enhancement of vibrational modes. Tip #1 is sensitive to both the HOMO and HOMO-1. Tip #2 is sensitive to none of the orbitals. Tips #3–5 are all sensitive to the HOMO, but with different resolution. Gray curves were acquired on the NaCl surface (denoted as "bkgd"). Reproduced with permission from Ref. [25]

e2 in the elastic process (Fig. 4.4d) should be enhanced and dominate over the other elastic channel, leading to the symmetric-to-asymmetric line shape change with decreasing tip height [22, 26, 34]. Those calculation results and analysis confirm that the observed IETS indeed involve a resonantly enhanced process due to the presence of the HOMO states around E_F, which are induced by the strong tip water coupling.

It is worth noting that the relative conductance change in the tip-enhanced IETS could be as high as 30%, thus facilitates the accurate determination of the H-bonding strength, which is related to the redshift of the D1 stretching mode [35]. Therefore, the development of tip-enhanced IETS provides a sensitive probe for NQEs of protons in energy space through isotope substitution experiments, which will be discussed in details in Chap. 6.

4.5 Action Spectroscopy

The instability of molecules induced by the perturbation of high energy inelastic tunneling electrons and propensity rule of IETS invokes one to think about another spectroscopic method which might be complementary to IETS. It is well known that vibrations of atoms or molecules may be a trigger for molecular diffusion or dissociation on surfaces, which can be directly monitored in STM by recording the tunneling current as a function of voltage bias [36–38]. Kawai et al. firstly coined

4.5 Action Spectroscopy

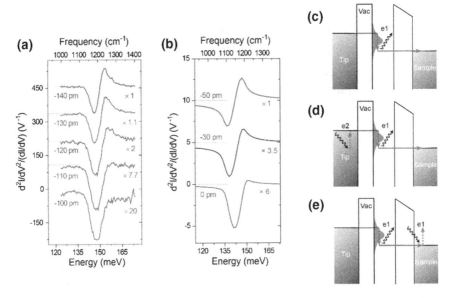

Fig. 4.4 Lineshape change of tip-enhanced IETS. a, b Experimental (**a**) and calculated (**b**) d^2I/dV^2 spectra (normalized by dI/dV) of the bending mode at different tip height. Tip heights in the experiment are referenced to the gap set with V = 100 mV and I = 50 pA. The tip height in calculation is defined as the vertical distance between the Cl atom at the tip apex and O atom of the water molecule. **c–e** Tunneling processes involved at the threshold bias and contributing to IETS. **c** Inelastic tunneling process. **d** Two-electron elastic process: One electron tunnels inelastically from the left to the right electrode, emitting a vibration; a second electron makes an inelastic transition upward at the left electrode, re-absorbing the emitted vibration. **e** One-electron elastic process: One electron makes an inelastic transition downward from the left to the right electrode, emitting a vibration; the same electron re-absorbs this vibration and makes an inelastic transition upward at the right electrode. Reproduced with permission from Ref. [25]

the word "action spectroscopy (AS)" in STM, to reveal the microscopic mechanisms of molecular dynamics and surface chemical reactions through electronic and vibrational excitation [39, 40]. In a typical STM-AS experiment, the reaction yield Y is measured as a function of sample bias. The observed AS would possibly show several thresholds related to the vibrational or electronic states responsible for the reaction. Single molecular rotation [41, 42], hopping [43], bond breaking [32] and conformational change [44] have been systematically and extensively explored using action spectroscopy.

Although limited by the inability to clarify all vibrational modes of a certain molecule and the requirement of an average of multiple measurements due to the intrinsic stochasticity of single-molecule reactions, STM-AS indeed holds complementary aspects compared with STM-IETS. It is widely recognized that IETS is governed by the propensity rule [45], where only some vibrational modes are detectable and predominantly contribute to the IET signals. Furthermore, competitions between elastic and inelastic channels in IETS signals complicated the line shape and vibrational energy assignment. One need to fit the asymmetric line shape spectra to extract the intrinsic vibrational energies. STM-AS, on the other hand, is directly related to the vibrational excitation. However, the reaction coordinate may be different from the excited vibrational mode, because the anharmonic coupling leads to energy transfer between different modes. Therefore, the interpretation of AS usually needs careful theoretical fitting [46]. Because of its statistical nature and the irreversibility of the chemical reactions, AS requires multiple measurements for a single action on different sites or even different molecules. In spite of the prolonged data-acquisition time, AS can provides deeper understanding of vibrationally mediated chemical reaction of single adsorbates, which is inaccessible from conventional IETS.

4.6 Summary

In this chapter, we have pushed the limit of vibrational spectroscopy of water down to the single-bond level using a novel technique called tip-enhanced (IETS) based on STM. This is achieved by gating the frontier orbitals of water towards the Fermi level with a chlorine-terminated STM tip to resonantly enhance the electron-vibration coupling. The signal-to-noise ratios of the tip-enhanced IETS are enhanced by orders of magnitude over the conventional STM-IETS, which is crucial for precisely determining the H-bonding strength and subsequently probing the NQEs of protons in energy space through isotope substitution experiments. More generally, the tip-enhanced IETS technique developed in this work defeats the long-standing deficiency of conventional IETS in the detection of weak intermolecular interaction, thus opens up a new avenue for studying water and other hydrogen-bonded systems.

References

1. Huisken F, Kaloudis M, Kulcke A (1996) Infrared spectroscopy of small size-selected water clusters. J Chem Phys 104:17–25
2. Buck U, Huisken F (2000) Infrared spectroscopy of size-selected water and methanol clusters. Chem Rev 100:3863–3890
3. Fecko CJ, Eaves JD, Loparo JJ, Tokmakoff A, Geissler PL (2003) Ultrafast hydrogen-bond dynamics in the infrared spectroscopy of water. Science 301:1698–1702
4. Bakker HJ, Skinner JL (2010) Vibrational spectroscopy as a probe of structure and dynamics in liquid water. Chem Rev 110:1498–1517
5. Bensebaa F, Ellis TH (1995) Water at surfaces: what can we learn from vibrational spectroscopy? Prog Surf Sci 50:173–185
6. Du Q, Superfine R, Freysz E, Shen YR (1993) Vibrational spectroscopy of water at the vapor/water interface. Phys Rev Lett 70:2313–2316
7. Shen YR, Ostroverkhov V (2006) Sum-frequency vibrational spectroscopy on water interfaces: polar orientation of water molecules at interfaces. Chem Rev 106:1140–1154
8. Kim Y, Motobayashi K, Frederiksen T, Ueba H, Kawai M (2015) Action spectroscopy for single-molecule reactions—Experiments and theory. Prog Surf Sci 90:85–143

References

9. Hirsch K, Holzapfel W (1986) Effect of high pressure on the Raman spectra of ice VIII and evidence for ice X. J Chem Phys 84:2771–2775
10. Goncharov AF, Struzhkin VV, Mao HK, Hemley RJ (1999) Raman spectroscopy of dense H_2O and the transition to symmetric hydrogen bonds. Phys Rev Lett 83:1998–2001
11. Nagata Y, Pool RE, Backus EHG, Bonn M (2012) Nuclear quantum effects affect bond orientation of water at the water-vapor interface. Phys Rev Lett 109:226101
12. Stipe BC, Rezaei MA, Ho W (1998) Single-molecule vibrational spectroscopy and microscopy. Science 280:1732–1735
13. Ho W (2002) Single-molecule chemistry. J Chem Phys 117:11033–11061
14. Komeda T (2005) Chemical identification and manipulation of molecules by vibrational excitation via inelastic tunneling process with scanning tunneling microscopy. Prog Surf Sci 78:41–85
15. Galperin M, Ratner MA, Nitzan A (2007) Molecular transport junctions: vibrational effects. J Phys: Condens Matter 19:103201
16. Lorente N, Persson M (2000) Theory of single molecule vibrational spectroscopy and microscopy. Phys Rev Lett 85:2997–3000
17. Galperin M, Ratner MA, Nitzan A, Troisi A (2008) Nuclear coupling and polarization in molecular transport junctions: beyond tunneling to function. Science 319:1056–1060
18. Morgenstern K, Nieminen J (2002) Intermolecular bond length of ice on Ag(111). Phys Rev Lett 88:066102
19. Kumagai T et al (2009) Tunneling dynamics of a hydroxyl group adsorbed on Cu(110). Phys Rev B 79:035423
20. Okuyama H, Hamada I (2011) Hydrogen-bond imaging and engineering with a scanning tunnelling microscope. J Phys D Appl Phys 44:464004
21. Kumagai T (2015) Direct observation and control of hydrogen-bond dynamics using low-temperature scanning tunneling microscopy. Prog Surf Sci 90:239–291
22. Persson BNJ, Baratoff A (1987) Inelastic electon tunneling from a metal tip: the contribution from resonant processes. Phys Rev Lett 59:339–342
23. Galperin M, Nitzan A, Ratner MA (2006) Resonant inelastic tunneling in molecular junctions. Phys Rev B 73:045314
24. Song H et al (2009) Observation of molecular orbital gating. Nature 462:1039–1043
25. Guo J et al (2016) Nuclear quantum effects of hydrogen bonds probed by tip-enhanced inelastic electron tunneling. Science 352:321–325
26. Lü JT et al (2014) Efficient calculation of inelastic vibration signals in electron transport: beyond the wide-band approximation. Phys Rev B 89:081405
27. Soler JM et al (2002) The SIESTA method for ab initio order-N materials simulation. J Phys Condens Matter 14:2745–2779
28. Brandbyge M, Mozos JL, Ordejon P, Taylor J, Stokbro K (2002) Density-functional method for nonequilibrium electron transport. Phys Rev B 65:165401
29. Frederiksen T, Paulsson M, Brandbyge M, Jauho A-P (2007) Inelastic transport theory from first principles: methodology and application to nanoscale devices. Phys Rev B 75:205413
30. Guo J et al (2014) Real-space imaging of interfacial water with submolecular resolution. Nat Mater 13:184–189
31. Lorente N, Persson M, Lauhon LJ, Ho W (2001) Symmetry selection rules for vibrationally inelastic tunneling. Phys Rev Lett 86:2593–2596
32. Ohara M, Kim Y, Yanagisawa S, Morikawa Y, Kawai M (2008) Role of molecular orbitals near the Fermi level in the excitation of vibrational modes of a single molecule at a scanning tunneling microscope junction. Phys Rev Lett 100:136104
33. Galperin M, Ratner MA, Nitzan A (2004) Inelastic electron tunneling spectroscopy in molecular junctions: peaks and dips. J Chem Phys 121:11965–11979
34. Baratoff A, Persson BNJ (1988) Theory of the local tunneling spectrum of a vibrating adsorbate. J Vac Sci Technol, A 6:331–335
35. Rozenberg M, Loewenschuss A, Marcus Y (2000) An empirical correlation between stretching vibration redshift and hydrogen bond length. Phys Chem Chem Phys 2:2699–2702

36. Stipe BC, Rezaei HA, Ho W (1999) Localization of inelastic tunneling and the determination of atomic-scale structure with chemical specificity. Phys Rev Lett 82:1724–1727
37. Stipe BC, Rezaei MA, Ho W (1998) Inducing and viewing the rotational motion of a single molecule. Science 279:1907–1909
38. Stipe BC et al (1997) Single-molecule dissociation by tunneling electrons. Phys Rev Lett 78:4410–4413
39. Kawai M, Komeda T, Kim Y, Sainoo Y, Katano S (2004) Single-molecule reactions and spectroscopy via vibrational excitation. Phil Trans R Soc Lond A 362:1163–1171
40. Kim Y, Komeda T, Kawai M (2002) Single-molecule reaction and characterization by vibrational excitation. Phys Rev Lett 89:126104
41. Sainoo Y et al (2005) Excitation of molecular vibrational modes with inelastic scanning tunneling microscopy processes: examination through action spectra of cis-2-butene on Pd(110). Phys Rev Lett 95:246102
42. Lauhon LJ, Ho W (2000) Electronic and vibrational excitation of single molecules with a scanning tunneling microscope. Surf Sci 451:219–225
43. Motobayashi K, Matsumoto C, Kim Y, Kawai M (2008) Vibrational study of water dimers on Pt(111) using a scanning tunneling microscope. Surf Sci 602:3136–3139
44. Kumagai T et al (2013) Thermally and vibrationally induced tautomerization of single porphycene molecules on a Cu(110) surface. Phys Rev Lett 111:246101
45. Paulsson M, Frederiksen T, Ueba H, Lorente N, Brandbyge M (2008) Unified description of inelastic propensity rules for electron transport through nanoscale junctions. Phys Rev Lett 100:226604
46. Frederiksen T, Paulsson M, Ueba H (2014) Theory of action spectroscopy for single-molecule reactions induced by vibrational excitations with STM. Phys Rev B 89:201302

Chapter 5
Concerted Proton Tunneling

5.1 Introduction

Proton transfer through hydrogen bonds plays an essential role in many physical, chemical and biological processes, such as phase transition, signal transduction, topological organic ferroelectrics, photosynthesis, and enzyme catalysis [1–9]. The complexity of proton dynamics largely arises from nuclear quantum effect in terms of proton tunneling. In comparison with the well-studied single proton tunneling, many-body tunneling is more complicated but participates in much broader proton dynamic processes, for instance, the phase transition of ice [10–12], molecular tautomerization and enzyme catalysis reactions [13–17]. However, our understanding of multiple proton tunneling is far from complete. Since protons within a H-bonding network are usually correlated [18], the proton tunneling may occur in a collective way. To date, some spectroscopic studies reported concerted proton tunneling in ices and H-bonded crystals [10, 11, 19], but the evidence is quite indirect and elusive. In addition, the correlated tunneling of protons is very sensitive to the local environment due to the demanding requirement of phase coherence among protons. Due to the limited spatial resolution, spectroscopic techniques measure the average properties of many H bonds, which could not provide such local information.

STM has shown the capability to probe and manipulate the intramolecular and intermolecular proton dynamics at the single-molecule level [20–26]. What is more, the impact of local environment on the proton dynamics has also been investigated by modifying the surrounding environment with atomic precision [25, 26]. However, the proton transfer dynamics in those works were mainly classical over-barrier hopping induced by inelastic tunneling electrons or thermal effects, whereas the through-barrier quantum tunneling of proton was seldom explored.

As a pioneering work, Lauhon and Ho directly observed with STM, the quantum tunneling of single H atoms on Cu(001) surface and further characterized the transition temperature (60 K) from thermally excited hopping to quantum tunneling based on the dependence of H/D hopping rate on the temperature [27]. Subsequently, the quantum tunneling of heavy atoms and molecules were also directly visualized with STM [28–30]. What's more, Kumagai et al. suggested that the H tunneling was involved in the flipping motion of OH/OD group and H bond exchange within the water dimer [21, 31]. The sequential proton tunneling has also been evidenced in the H-bonded water-hydroxyl chain [24]. Recently, Koch et al. demonstrated the direct observation of stepwise double proton tunneling in a single porphycene molecule on a Ag(110) Surface [17]. In spite of those great achievements, whether the concerted proton tunneling exists or not still remains controversial [11, 12, 32, 33]. In this section, we present the direct visualization of concerted proton tunneling in a cyclic water tetramer adsorbed on the NaCl(001) surface [34, 35] using the submolecular-resolution imaging technique [36].

5.2 Chirality Switching of Water Tetramers

Water tetramers adsorbed on the NaCl surface contains two different chiral states, clockwise and anticlockwise H-bonded loops, which can be distinguished by orbital imaging based on STM [36] or non-invasive Δf imaging with nc-AFM [37]. Obviously, these two different chiral states could be switched once the four H-boned protons transfer to the nearest oxygen atoms collectively. We show the possibility of manipulating the reversible interconversion of two different chiral states of the water tetramer in a well-controlled manner with a Cl-terminated tip (Fig. 5.1a), which can provide a long-range electrical interaction with the protons owing to its electronegative nature.

The switching dynamics of the tetramer chirality can be monitored by recording tunneling current as a function of time (Fig. 5.1b). The Cl-terminated tip is positioned slightly off the center of a clockwise tetramer (marked as green stars in inset of Fig. 5.1b) in order to read out the current difference between the two chiral states. As shown in Fig. 5.1, the water tetramer stays at the clockwise state when the tip is far away. Once the Cl-tip is approached toward the tetramer with the distance of 230 pm, the tunneling current increases immediately along with splitting into two levels. The upward and downward jump of the tunneling current from one level to the other arises from the reversible interconversion between the two chiral states. Retracting the tip to the initial height at any of the two levels left the tetramer in a certain chirality, which could be identified by high-resolution orbital imaging. This method allows us to attribute the high and low current level to the anticlockwise state (AS) and clockwise state (CS) of the water tetramer, respectively.

Besides the two-level current trace, multi-level (three, four, five and six levels) current traces are also observed, which may arise from the structural relaxation of the Cl atom adsorbed at the tip apex at small tip-water distance [35]. If the Cl atom is firmly absorbed on the tip and the shape of the tip apex is regular, there are only two levels in the current trace as shown in Fig. 5.1b. We noticed that no matter how many levels are there, they could be divided into two groups and each group corresponds to one chiral state of the water tetramer. In order to analyze the chirality switching

5.2 Chirality Switching of Water Tetramers

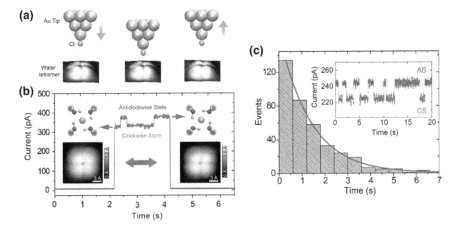

Fig. 5.1 Chirality switching of water tetramer. a Manipulation of chirality switching of water tetramer by decreasing the tip height. **b** Tunneling current trace during chirality switching acquired at the position of green stars on the water tetramer. **c** Extraction of the chirality switching rate by fitting the lifetime distribution to an exponential decay with a time constant of 1.37 s. The switching rate is the inverse of the time constant, (0.73 ± 0.016) s^{-1}. Reproduced with permission from Ref. [35]

in a quantitative way, the switching rates of AS → CS and CS → AS are extracted from the current versus time trace by fitting the lifetime distribution of the higher and lower current levels, respectively, to an exponential decay $y = Ae^{-t/\tau}$ (red curve in Fig. 5.1c). The switching rate was obtained simply by taking the inverse of the time constant τ. To obtain reliable statistics of the intervals of each state, hundreds of switching events are required. In the cases where these two current levels are so close that the separation is comparable to the noise background, adjacent averaging method is adopted to restore the two current levels.

5.3 Quantitative Analysis of the Switching Rate

5.3.1 Impact of Bias on Switching Rate

In order to explore the mechanism of proton transfer in the water tetramer, we measure the switching rate as a function of bias voltage (Fig. 5.2a). The switching rates of AS → CS and CS → AS remain nearly the same within the sample bias range from −6 to 4 mV, suggesting that both of the switching rates are independent on the magnitude as well as the polarity of the bias voltage and the tunneling current. It is worth noting that the switching dynamics did not cease near the zero bias, but with the same rate as the case of finite bias voltages. These results demonstrate that the proton transfer in the water tetramer is neither induced by the excitation of the

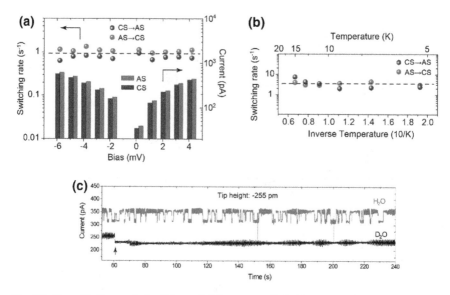

Fig. 5.2 Quantitative analysis of the switching rate. a Switching rate as functions of sample bias and tunneling current. **b** Temperature dependence of switching rate. **c** Isotope effect on the switching rate. Reproduced with permission from Ref. [35]

inelastic tunneling electrons [38, 39], nor by the electric field between the tip and the sample [40], though chirality switching by inelastic electron tunneling has been observed previously [39].

5.3.2 Dependence of Switching Rate on Temperature

We carry out tetramer chirality switching experiment at different temperature to investigate the thermal effect. Upon changing the temperature from 5 to 15 K, the switching rates of the water tetramer stay almost constant except for the slight increase at 15 K (Fig. 5.2b), clearly deviating from the Arrhenius-law behavior, which excluded the influence of the thermal fluctuations [27]. Therefore, it is very likely that the proton transfer in the water tetramer is driven by quantum tunneling.

5.3.3 Isotope Effect of Switching Rate

To confirm the proton transfer in tetramer is dominated by proton tunneling, we perform isotope substitution experiment. As shown in Fig. 5.2c, the switching rate of a D_2O tetramer dropped two orders of magnitude with respect to the H_2O tetramer

(Fig. 5.2c). Therefore, we attribute the interconversion between the two chiral states to quantum tunneling of protons between the water molecules. The similar isotope effect has been observed in the H bond exchange process within a water dimer adsorbed on the Cu(110) surface at 6 K [21]. However, whether the many-body tunneling is concerted or sequential is still unclear.

Bove et al. found that the disappearance of experimental signal in the related neutron-scattering study upon partial deuteration [11], thus suggested the concerted proton tunneling in ice. This observation is further confirmed and explained theoretically [33]. Partial deuteration of the H system leads to inequivalent protons, thereby destroying concerted proton tunneling. Inspired by these works, we carry out partial isotope substitution on the chirality switching to explore the tunneling of the four protons is concerted or not.

The partial deuterated water tetramer ($3H_2O+D_2O$) is constructed by manipulating four individual water molecules (three H_2O and one D_2O monomer) with the Cl-terminated tip (Fig. 5.3a–f). The H_2O and D_2O molecules could be distinguished by reference to the frequency of the O–H/D stretching mode (Fig. 5.3g) with the recently developed tip-enhanced IETS. The formation of $3H_2O+D_2O$ tetramer can be evidenced by the characteristic OH and OD vibrational features in the d^2I/dV^2 spectrum of the water tetramer, as shown in the green curve of Fig. 5.3g. Strikingly, the chirality switching rate of $4H_2O$ tetramer is substantially reduced once replacing only one H_2O with D_2O, almost to the same level of $4D_2O$ tetramer (Fig. 5.3h). This is in qualitative agreement with the prediction of the breakdown of concerted tunneling, as induced by a Zundel-like complex formed at the transition state (TS) upon partial isotopic substitute of H by D in ice I_h [33]. Therefore, we view this rapid quenching of the chirality switching of the partial deuterated water tetramer ($3H_2O+D_2O$) as a strong experimental support for the concerted proton tunneling in H_2O tetramer.

5.3.4 Energy Profiles of Chirality Switching in Water Tetramers

For a better understanding of the atomistic details of the chirality switching between CS and AS, we now resort to ab initio theoretical simulations. At the atomic scale, the chirality switching of water tetramer can occur via different channels. Stepwise proton transfer, intuitively the simplest process is investigated using the cNEB method, showing a high reaction barrier (>2.0 eV) (Fig. 5.4) duo to the formation of charged defects. Therefore, we rule out stepwise transfer mode and focus on the concerted transfer and rotation switching channels.

We propose four switching channels, including collective proton transfer (inset of Fig. 5.5a), collective water molecules rotation involving one OH bond (inset of Fig. 5.5b), two OH bonds (inset of Fig. 5.5c), and stepwise water molecules rotation (inset of Fig. 5.5d). In comparison with the other switching channels, the

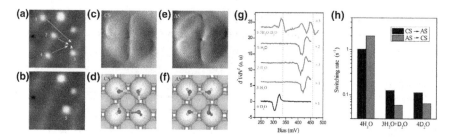

Fig. 5.3 Effect of full/partial isotope substitution on the chirality switching of water tetramer. **a, b** The construction of water tetramer by manipulating three H_2O monomers (1, 2, 3) and one D_2O monomer (4) with a Cl-terminated tip. **c–f** Derivative STM images and configuration model of the water tetramers with clockwise state (CS) and anti-clockwise state (AS) H-bonded loops. H, D, O, Cl^- and Na^+ are denoted by pink, blue, red, green and yellow spheres, respectively. **g** The d^2I/dV^2 spectra of the H_2O/D_2O monomers and partially deuterated water tetramer composed by three H_2O and one D_2O. Number 1–4 means the spectra of the water monomers in **a**. The d^2I/dV^2 spectrum of the constructed $3H_2O+D_2O$ tetramer is shown in green curve. The vertical dashed lines denote the vibrational energies of the free OD (338 meV) and OH (460 meV) stretching modes. **h** Switching rates of the water $4H_2O$ tetramer, $3H_2O+D_2O$ tetramer and $4D_2O$ tetramer. Reproduced from Ref. [41] with permission

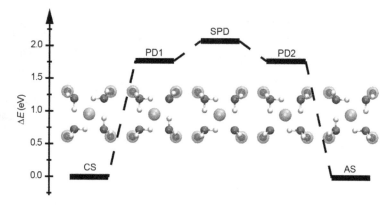

Fig. 5.4 Energy profile of the stepwise proton transfer in a water tetramer. In a stepwise proton transfer process two paired defect states (PD1 and PD2) and a separated paired state (SPD) are visited. Insets show the structures of these states. Reproduced with permission from Ref. [35]

climbing image nudged elastic band (cNEB) barrier for the stepwise rotation pathway is the lowest (Fig. 5.5d). Now we investigate the NQEs on other chirality switching channels by comparing the energy profiles from cNEB calculations (black dots) and path-integral molecular dynamics (PIMD) simulations (Fig. 5.5e–h). In contrast to the substantial reduction of the switching barrier for collective proton transfer, NQEs on the molecule rotation channels are much weaker. This is explained by plotting distribution probabilities of the path-integral images as a function of the reaction coordinates (Φ^T and Φ^R) at the TSs (purple shades in Fig. 5.5e–h). For collective proton transfer, obvious proton delocalization (Fig. 5.5e) between the two degenerate

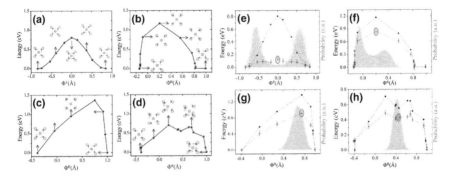

Fig. 5.5 Reaction barrier of chirality switching. a–d Energy profiles of the chirality switching in water tetramer via **a** the four-proton concerted tunneling, the collective rotations of four water molecules involving **b** one and **c** two OH bonds, respectively, **d** the step-wise rotation of four water molecules. **e–h** Energy profiles from cNEB calculations (black dots) and PIMD simulations (red triangles), for four kinds of water chirality switching modes as shown in **a–d**. The distribution probabilities of reaction coordinates are also shown by shades in each figure for the TSs. Reproduced with permission from Ref. [41]

states along H-bonds can be seen and deep tunneling plays an important role at 50 K. However, this delocalization is much weaker in rotation modes (Fig. 5.5f) or even does not exist at all (Fig. 5.5g, h). In summary, the concerted proton tunneling in water tetramer is confirmed by presenting a much smaller free-energy barrier for the translational collective proton tunneling mode than other chirality switching modes at low T.

5.4 Impact of Local Environments on Concerted Proton Tunneling

The Cl-tip can be used to probe the impact of atomic-scale environment on the concerted proton tunneling by tuning the coupling between the Cl-tip and the water tetramer in xyz dimensions. We first investigate the effect of tip height (z dimension). The proton tunneling in the tetramer shows a distinct dependence on the tip height (Fig. 5.6a). The switching rates first undergo an initial rise (region I) followed by a rapid drop (region II) with decreasing tip height. We notice that the switching rate of CS → AS is much larger than the AS → CS at small tip height due to the asymmetric double-well potential, thus the water tetramer prefers to stay at the AS. Such an asymmetry can be understood considering the fact that the STM tip was not located at the perfect geometry center of the water tetramer so as to distinguish the two chiral states.

In order to understand the dependence of switching rate on the tip height, we calculated the effective energy barrier by subtracting from the original potential

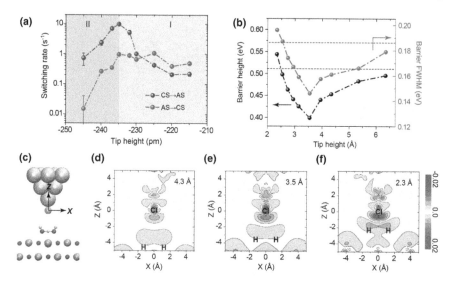

Fig. 5.6 Dependence of the switching rate on the tip height. a Switching rate as a function of tip height. **b** Calculated effective barrier height and full-width at half-maximum (FWHM) as a function of tip height after considering the ZPE difference between the initial and transition states. The dashed lines denote the barrier height and width without the tip. **c** Schematic model of Cl-tip/water tetramer/NaCl system. **d–f** Calculated electron density difference of the experimental system **a** at the plane perpendicular to the NaCl surface with different tip-water distance. Red and blue in the color bar represent electron gain and depletion, respectively. The units of electron density are e Å^{-3}. Reproduced with permission from Ref. [35]

barrier the zero-point energy difference between the initial and transition states of the adsorbed water tetramer (Fig. 5.6b). As the tip moved towards the water tetramer, both the barrier height and width fall off. Further decreasing the tip height, the barrier height and width exhibit a reversal behavior, which is consistent with the experimental observations (Fig. 5.6a). We have also calculated the energy barriers of tetramer switching via the concerted rotation of four water molecules, in which both the barrier height and width (FWHM) decrease monotonically as the tip height decreases below 3.5 Å [35]. Following concerted rotation manner, the tunneling rate should be enhanced very quickly at small tip height. This behavior is qualitatively different from the experimental observation of the tip height dependence on the tunneling rate (region II in Fig. 5.6a), which shows a rapid drop as tip approaches the tetramer, thus excluding the rotation pathway during the chirality switching of the tetramer.

To clarify the crucial role of the Cl-decorated tip in the tetramer switching, we calculated the electron density differences of Cl-tip/tetramer/NaCl-bilayer system (Fig. 5.6d–f) in a plane perpendicular to the surface (Fig. 5.6c). DFT calculations reveal that the Cl atom at the tip apex is not neutral, but negatively charged with a partial charge of 0.4e. When a Cl-tip was placed above the tetramer, local charge rearrangement appeared with depletion (accumulation) of density on the H (Cl)

5.4 Impact of Local Environments on Concerted Proton Tunneling

atoms, implying the formation of weak electrostatic attractive interaction between the H^+ and $Cl^{\delta-}$ (Fig. 5.6d). At smaller tip height, the charge rearrangement is enhanced producing stronger electrostatic force (Fig. 5.6e), which facilitates the proton transfer through the H bonds, resulting to a smaller reaction barrier. However, when the tip is too close to the water molecule, the coulomb repulsion between the O^{2-} and $Cl^{\delta-}$ gradually dominants over the H^+ and $Cl^{\delta-}$ attraction interaction (Fig. 5.6f), leading to the larger O–O distances and consequently the rapid increase of reaction barrier. Therefore, it is the delicate competition between the H^+–$Cl^{\delta-}$ attraction and the O^{2-}–$Cl^{\delta-}$ repulsion that leads to an initial drop of reaction barrier followed by a rapid rise.

Another very interesting finding is that the concerted proton tunneling is extremely sensitive to the lateral position of the Cl-tip (x–y dimensions). Moving the tip only 0.5 Å from the center to the edge of the tetramer, the switching rates dropped by almost one order of magnitude (Fig. 5.7). Such a fast decay is closely related to the collective and concerted nature of the proton tunneling. When the Cl-tip was at the center of the tetramer, the Cl anion at the tip apex interacted with the four protons equally. Then four protons could be considered as a quantum quasiparticle and moved in a fully correlated manner. However, the degeneracy of the four H bonds was broken because of the asymmetric coupling between the Cl anion and the four protons once the tip was positioned off the center of the tetramer. As a result, the correlated proton tunneling would be dramatically suppressed.

These systematic studies demonstrate the key role of local environment in the concerted tunneling process, showing that the presence of the Cl anion at the tip apex may either enhance or suppress the concerted tunneling process in water tetramers, depending on the details of the coupling symmetry between the Cl ion and the protons. This provides us the opportunity to continuously suppress the reaction barrier

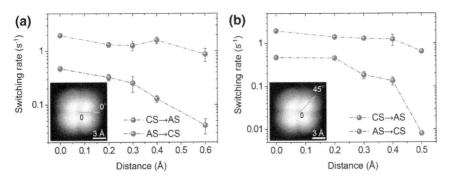

Fig. 5.7 Dependence of the switching rate on the lateral position of the Cl-tip. a, b Switching rates acquired at the different lateral positions of the tip along the 0° and 45° direction away from the center of the tetramer (highlighted by the green arrows), respectively. The zero point of the tip position is set about 0.1 Å away from the center of the tetramer so as to distinguish the two different chiral states in the tunneling current. The error bars represent the standard error. Reproduced with permission from Ref. [35]

such that the reaction barrier may even lower than the zero-point motion, leading to the quantum delocalization of the proton, by proper engineering the tip apex with different terminations and precise controlling the tip-water coupling.

5.5 Summary

The concerted proton tunneling in a water tetramer is directly visualized by monitoring the reversible interconversion of the H-bonding directionality and associated chirality of the tetramer using submoleuclar orbital-imaging technique. We also show the possibility of controlling the quantum states of protons with atomic-scale precision via tuning the coupling between the Cl-tip and the water tetramer.

Recently, similar concerted proton tunneling has been also observed in water hexamer prism and bulk ice, using rotational spectroscopy [19] and dielectric measurement [42], respectively. Therefore, it may be a very general phenomenon in water/ice and other H-bonded materials. In addition, the existence of concerted proton tunneling implies that the proton dynamics in ice does not freeze out even at very low temperature [11], which may be responsible for the nonzero value of the zero-temperature entropy of ices I_h [43]. A recent theoretical study pointed out that the protons in ice I_h could form a quantum liquid at low temperatures, in which protons are not merely disordered, but continually fluctuate between different configurations obeying the ice rules [44]. More strikingly, water confined inside a hexagonal shaped channel of the gemstone beryl exhibited a new quantum state because of quantum tunneling and delocalization [45]. As a result, the quantum nature of proton plays a significant effect on the structure, dynamics and properties of water and ice.

References

1. Löwdin P-O (1963) Proton tunneling in DNA and its biological implications. Rev Mod Phys 35:724–732
2. Caldin EF (1969) Tunneling in proton-transfer reactions in solution. Chem Rev 69:135–156
3. Kreuer KD (1996) Proton conductivity: materials and applications. Chem Mater 8:610–641
4. Benoit M, Marx D, Parrinello M (1998) Tunnelling and zero-point motion in high-pressure ice. Nature 392:258–261
5. Frank RAW, Titman CM, Pratap JV, Luisi BF, Perham RN (2004) A molecular switch and proton wire synchronize the active sites in thiamine enzymes. Science 306:872–876
6. Marx D (2006) Proton transfer 200 years after von Grotthuss: insights from ab initio simulations. ChemPhysChem 7:1848–1870
7. Masgrau L et al (2006) Atomic description of an enzyme reaction dominated by proton tunneling. Science 312:237–241
8. Horiuchi S et al (2010) Above-room-temperature ferroelectricity in a single-component molecular crystal. Nature 463:789–792
9. Marx D, Chandra A, Tuckerman ME (2010) Aqueous basic solutions: hydroxide solvation, structural diffusion, and comparison to the hydrated proton. Chem Rev 110:2174–2216

10. Brougham DF, Caciuffo R, Horsewill AJ (1999) Coordinated proton tunnelling in a cyclic network of four hydrogen bonds in the solid state. Nature 397:241–243
11. Bove LE, Klotz S, Paciaroni A, Sacchetti F (2009) Anomalous proton dynamics in ice at low temperatures. Phys Rev Lett 103:165901
12. Lin L, Morrone JA, Car R (2011) Correlated tunneling in hydrogen bonds. J Stat Phys 145:365–384
13. Douhal A, Kim SK, Zewail AH (1995) Femtosecond molecular dynamics of tautomerization in model base pairs. Nature 378:260–263
14. Billeter SR, Webb SP, Agarwal PK, Iordanov T, Hammes-Schiffer S (2001) Hydride transfer in liver alcohol dehydrogenase: quantum dynamics, kinetic isotope effects, and role of enzyme motion. J Am Chem Soc 123:11262–11272
15. Pu JZ, Gao JL, Truhlar DG (2006) Multidimensional tunneling, recrossing, and the transmission coefficient for enzymatic reactions. Chem Rev 106:3140–3169
16. Kwon OH, Zewail AH (2007) Double proton transfer dynamics of model DNA base pairs in the condensed phase. Proc Natl Acad Sci USA 104:8703–8708
17. Koch M et al (2017) Direct observation of double hydrogen transfer via quantum tunneling in a single porphycene molecule on a Ag(110) surface. J Am Chem Soc 139:12681–12687
18. Castro Neto AH, Pujol P, Fradkin E (2006) Ice: a strongly correlated proton system. Phys Rev B 74:024302
19. Richardson JO et al (2016) Concerted hydrogen-bond breaking by quantum tunneling in the water hexamer prism. Science 351:1310–1313
20. Liljeroth P, Repp J, Meyer G (2007) Current-induced hydrogen tautomerization and conductance switching of naphthalocyanine molecules. Science 317:1203–1206
21. Kumagai T et al (2008) Direct observation of hydrogen-bond exchange within a single water dimer. Phys Rev Lett 100:166101
22. Kumagai T et al (2010) Symmetric hydrogen bond in a water-hydroxyl complex on Cu(110). Phys Rev B 81:045402
23. Auwarter W et al (2012) A surface-anchored molecular four-level conductance switch based on single proton transfer. Nat Nanotech 7:41–46
24. Kumagai T et al (2012) H-atom relay reactions in real space. Nat Mater 11:167–172
25. Kumagai T et al (2014) Controlling intramolecular hydrogen transfer in a porphycene molecule with single atoms or molecules located nearby. Nat Chem 6:41–46
26. Kumagai T (2015) Direct observation and control of hydrogen-bond dynamics using low-temperature scanning tunneling microscopy. Prog Surf Sci 90:239–291
27. Lauhon LJ, Ho W (2000) Direct observation of the quantum tunneling of single hydrogen atoms with a scanning tunneling microscope. Phys Rev Lett 85:4566–4569
28. Heinrich AJ, Lutz CP, Gupta JA, Eigler DM (2002) Molecule cascades. Science 298:1381–1387
29. Repp J, Meyer G, Rieder KH, Hyldgaard P (2003) Site determination and thermally assisted tunneling in homogenous nucleation. Phys Rev Lett 91:206102
30. Stroscio JA, Celotta RJ (2004) Controlling the dynamics of a single atom in lateral atom manipulation. Science 306:242–247
31. Kumagai T et al (2009) Tunneling dynamics of a hydroxyl group adsorbed on Cu(110). Phys Rev B 79:035423
32. Drechsel-Grau C, Marx D (2014) Quantum simulation of collective proton tunneling in hexagonal ice crystals. Phys Rev Lett 112:148302
33. Drechsel-Grau C, Marx D (2014) Exceptional isotopic-substitution effect: breakdown of collective proton tunneling in hexagonal ice due to partial deuteration. Angew Chem Int Ed 53:10937–10940
34. Drechsel-Grau C, Marx D (2015) Tunneling in chiral water clusters. Protons in concert. Nat Phys 11:216–218
35. Meng X et al (2015) Direct visualization of concerted proton tunnelling in a water nanocluster. Nat Phys 11:235–239
36. Guo J et al (2014) Real-space imaging of interfacial water with submolecular resolution. Nat Mater 13:184–189

37. Peng JB et al (2018) Weakly perturbative imaging of interfacial water with submolecular resolution by atomic force microscopy. Nat Commun 9:112
38. Ho W (2002) Single-molecule chemistry. J Chem Phys 117:11033–11061
39. Huang T et al (2011) A molecular switch based on current-driven rotation of an encapsulated cluster within a fullerene cage. Nano Lett 11:5327–5332
40. Gawronski H, Carrasco J, Michaelides A, Morgenstern K (2008) Manipulation and control of hydrogen bond dynamics in absorbed ice nanoclusters. Phys Rev Lett 101:136102
41. Feng YX et al (2018) The collective and quantum nature of proton transfer in the cyclic water tetramer on NaCl(001). J Chem Phys 148:102329
42. Yen F, Gao T (2015) Dielectric anomaly in ice near 20 K: evidence of macroscopic quantum phenomena. J Phys Chem Lett 6:2822–2825
43. Pauling L (1935) The structure and entropy of ice and of other crystals with some randomness of atomic arrangement. J Am Chem Soc 57:2680–2684
44. Benton O, Sikora O, Shannon N (2016) Classical and quantum theories of proton disorder in hexagonal water ice. Phys Rev B 93:125143
45. Kolesnikov AI et al (2016) Quantum tunneling of water in beryl: a new state of the water molecule. Phys Rev Lett 116:167802

Chapter 6
Nuclear Quantum Effect of Hydrogen Bonds

6.1 Introduction

Hydrogen-bonding interaction is ubiquitous in nature and plays an essential role in a broad spectrum of physics, chemistry, biology, energy and material sciences. NQEs, in terms of zero-point fluctuation, could influence the H-bonding interactions and consequently the structure of H-bonded networks due to the anharmonic nature of the potential well (Fig. 1.5b). For instance, high-pressure ice exhibits a prominent proton delocalization effect between the oxygen atoms due to the relatively small O–O distance, leading to the blurring between covalent bond and H bond [1–4]. Further PIMD simulations revealed that the magnitude of proton delocalization is quite sensitive to the distance of two nearest neighbouring oxygen atoms, as evidenced in the water-hydroxyl complex on metal surfaces [5]. Experimentally, the proton delocalization effect is confirmed in interfacial water, where symmetric H bond (HO–H–OH) was visualized in a water-hydroxyl complex on Cu(110) surface by STM [6].

Conventionally, the Ubbelohde effect yields an elongation of the O–O distance and the weakening of H bond upon deuteration, whereas the reverse Ubbelohde effect has also been observed [7, 8]. Based on the SFG spectroscopy technique and PIMD calculations method, Nagata et al. revealed that the NQEs influence the bond orientation of the interfacial water (HOD), in which the OH bond prefers to orient up toward the vapor phase, whereas the OD tends to form a D bond with the bulk water because of the stronger D bond of the deuterium [9]. In HF H-bonded systems, both weakening and strengthening of H bond by NQEs was reported by the theoretical PIMD calculations, depending on the cluster size [10–12]. In spite of the enormous theoretical efforts devoted to proper treatment of the nuclear motion at the quantum mechanical level [13–17], a clear cohesive picture for the influence of NQEs on the H-bonding strength at the atomic level is still not available and it still remains an open question how large the quantum component of the H bond is.

As described in Chap. 4, the recently developed tip-enhanced IETS enables us to probe the vibrational energy of O–H/D stretching mode with ultrahigh signal-to-noise ratio. Then, the H-bonding strength could be estimated from the redshift (softening) in the X-H stretching frequency of the H-bond donor molecule [18]. In this chapter, I will report the quantitative assessment of NQEs on the strength of a single H bond formed at a water-salt interface using the tip-enhanced IETS technique. Then I explore the key impact of the local environment on the NQEs and unravel the background physical picture based on the PIMD simulations.

6.2 Measurement of H-Bonding Strength by Tip-Enhanced IETS

6.2.1 Tip-Enhanced IETS of Stretching Mode

We focus on the stretching mode because it is most sensitive to the H-bonding interactions. However, measuring reliable signals of stretching mode with STM-IETS has long been a challenge. Using tip-enhanced IETS, we can obtain the high resolution stretching band vibrational energies of water monomer (Fig. 6.1). As shown in Fig. 4.1a, water monomers adsorbed on the Au(111)-supported NaCl(001) film in a "standing" configuration [19] with one OD bond (D2) of a D_2O molecule dangles upward, and the other (D1) forms a H-bond with the Cl^- of NaCl surface. Accordingly, the zoom-in high resolution IETS of the adsorbed D_2O shows two distinct dip features, which are highlighted as D1 and D2 in Fig. 6.1a, respectively. The energy of D2 nearly coincides with that of the free OD stretching, and thus should mainly arise from the upright dangling OD bond. The other mode (D1), which was considerably red-shifted, we attributed to the downward OD H-bonded with the NaCl surface. For H_2O, only the red-shifted mode (H1) could be observable, the higher mode (H2) was too weak to show any detectable signal (Fig. 6.1a). The measurement of the stretching mode of H_2O monomer is more difficult than D_2O because of the larger vibration energy, and subsequently higher perturbation probability.

We notice that the symmetric stretching mode shows much larger signal than the asymmetric stretching mode. This can be understood from the symmetry of the vibrational modes and electronic states. Both LUMO and HOMO orbitals are symmetric with respect to the plane containing the line equally dividing the D–O–D angle, and perpendicular to the D–O–D plane (Fig. 6.1b). In turn, the electronic scattering states involved in the vibration-assisted transition approximately follow this symmetry. Since asymmetric stretching mode is anti-symmetric about this plane, its matrix element should be much smaller than that of the symmetric stretching mode, as shown in Fig. 6.1b. The selective excitation of the vibrational modes was also observed in action spectroscopy, which depends on the spatial distribution and the population of the molecular orbitals near the E_F [21].

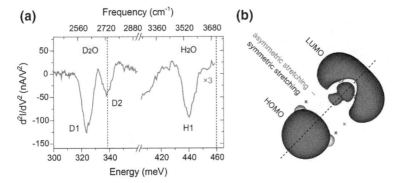

Fig. 6.1 High-resolution IETS of stretching mode of water monomers. a d^2I/dV^2 spectra of D_2O (blue) and H_2O (red) monomers. Vertical dashed lines denote the vibrational energies of the free OD and OH stretching modes, 338 meV and 460 meV, respectively. **b** The schematic of symmetry of electronic states (HOMO/LUMO) and vibrational modes. Reproduced with permission from Ref. [20]

6.2.2 Tuning of H-Bonding Strength

By adjusting the tip height, we can continuously modify the strength of the H bond formed between the water and the NaCl substrate in analogy to high-pressure experiments. The two-dimensional (2D) color map of the stretching bands of a D_2O monomer as a function of tip height (Fig. 6.2a) shows that both D1 and D2 are red-shifted with decreasing tip height. D2 mode vanishes at a certain tip height (marked by two arrows), where D1 mode changes from symmetric line shape to asymmetric Fano-shape feature. What's more, the vibrational energy of D1 mode keeps red-shifting when the tip is further approaching to the water monomer. In a word, with decreasing tip height, the stretching modes move towards lower energies, suggesting the enhancement of the H-bonding strength.

In order to confirm such a manner, we calculated the electron density difference of the whole tip/water/NaCl system (Fig. 6.2b–d). When the tip is absent (Fig. 6.2b), it can be seen that there is no obvious charge transfer or orbital hybridization between the water and NaCl but local charge rearrangement. Therefore, the water-NaCl interaction cannot be strong covalent or ionic bonding. The nature of the charge rearrangement between the D1 and the Cl^- is characteristic of the H-bonding with depletion (accumulation) of density on the D1 (Cl^-) atoms implicated in the H-bonds, which is highlighted by a dashed ellipse (Fig. 6.2b). The charge rearrangement between the O and the Na^+ arises mostly from their electrostatic attractive interaction. Calculation results show that the Cl atom at the tip apex is negatively charged, with a partial charge of about 0.4e. When the Cl-tip is introduced, a H-bond also forms between the water and the tip as suggested by the local charge rearrangement (highlighted by the upper ellipse in Fig. 6.2c).

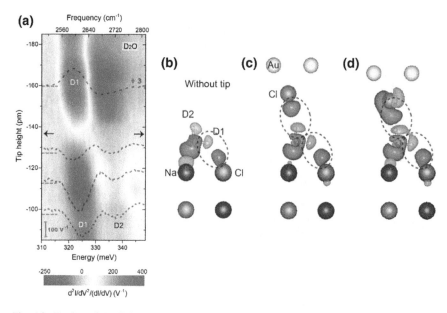

Fig. 6.2 Tuning of the H-bonding strength. a Two dimensional (2D) IETS color map of the D1 and D2 mode of a D_2O monomer as a function of tip height. Four typical IET spectra at different tip heights (dashed curves) are overlaid onto the color map. **b–d** The isosurface contours of differential charge density of the water/NaCl system without **b** and with Cl-terminated tip at the tip height of 4.25 Å **c** and 3.45 Å **d**. Red and blue represent charge gain and depletion, respectively. Dashed ellipses denote the local charge arrangement due to the water-NaCl interactions and tip-water interactions. Reproduced with permission from Ref. [20]

As the tip is further approached to the water (Fig. 6.2d), the water is gradually pushed close to the NaCl due to the Pauli repulsion force between the close-shell water molecule and the Cl-tip. The resulting changes of the differential charge density mainly reside in the regions of two H-bonds (O-D1···Cl^- and O-D2···Cl-tip), where more electron accumulation and depletion are observed. This is a clear evidence of the strengthening of the two H-bonds, leading to the measured red shifts of the D1 and D2 stretching modes (Fig. 6.2a). It is worthy to be noted that both the tip-water interaction and the water-NaCl interaction are quite local. Therefore, the stretching mode of the upper (lower) OD bond is primarily sensitive to the tip-water (water-NaCl) interaction. Due to the high quality of the IETS data, we are able to simultaneously resolve those two O–D stretching frequencies within the same D_2O monomer (Fig. 6.1a), such that the tip-water interaction and the water-NaCl interaction are unambiguously separated.

6.2 Measurement of H-Bonding Strength by Tip-Enhanced IETS

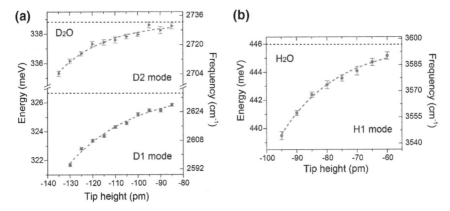

Fig. 6.3 Extraction of the intrinsic vibrational energies. Energies of D1, D2 **a** and H1 **b** stretching modes as a function of tip height. Each group of data is fitted to an inversed exponential decay. Extrapolating these curves to infinite tip height allows us to obtain the intrinsic energies (dashed lines) without the influence of the tip. Reproduced with permission from Ref. [20]

6.2.3 Extraction of the Intrinsic Vibrational Energies

In order to eliminate the influence of the tip-water coupling, we plot the energies of D1, D2 and H1 modes as a function of tip height, which can be fitted to inversed exponential decays (Fig. 6.3). Extrapolating these curves to infinite tip height allows us to extract the intrinsic energies of the vibrational modes. The resulting intrinsic energies of D2, D1 and H1 stretching modes are 338.7 ± 0.3, 326.8 ± 0.5 and 446 ± 0.4 meV, respectively. Compared with the free OD stretching mode (338 meV), the extrapolated D2 energy (338.7 ± 0.3 meV) is mildly blue-shifted, mainly arising from the intramolecular vibrational coupling between the D1 and D2 transition dipoles [22].

6.3 Impact of NQEs on the Strength of a Single H Bond

6.3.1 Tip-Enhanced IETS of HOD Monomers

As we have already discussed in the last section, the intramolecular vibrational coupling between D1 and D2 transition dipoles of the D_2O monomer leads to an additional shift of the measured energies, thus complicating the determination of the H-bonding energy. However, we notice that the intramolecular coupling is not exist in HOD monomers because of the large energy mismatch of the OH and OD stretching mode. So, we measure the IETS of the HOD monomers. Surprisingly, both the OH and OD stretching modes show considerable redshifts (Fig. 6.4a), suggesting they are both H-bonded. At smaller tip heights, the D2 mode appears coexisting with

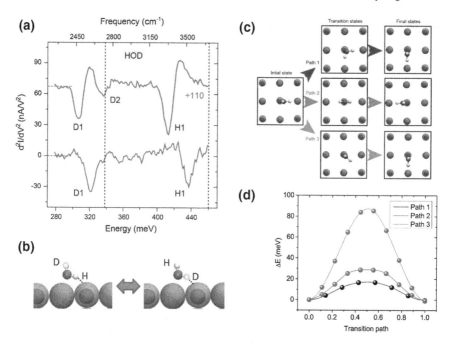

Fig. 6.4 Tip-enhanced IETS of a HOD monomer. a IET spectra of the stretching mode of a HOD monomer at the tip height of −1.4 Å (blue curve) and −2.3 Å (red curve). **b** Schematic model of the alternatively formation of H bond between the OH/OD and the Cl⁻ via rapidly flipping of the HOD monomer. O, H, D, Cl⁻ and Na⁺ are denoted by red, white, golden, grey and dark-cyan spheres, respectively. **c** Top views of the initial, transition and final states along three different paths for the flipping motion. **d** Nudged elastic band (NEB) calculations of energy profiles of water monomer flipping. Reproduced with permission from Ref. [20]

the D1 and H1 mode (Fig. 6.4a). These features indicate that the water monomer might be rapidly flipping so that the OH and OD form H-bonds with the NaCl surface alternatively (Fig. 6.4b). To confirm such a flipping behavior, we calculated the flipping barrier using NEB method in VASP. We found that the energy barrier for the flipping motion is as small as 20 mV [20] (Fig. 6.4c, d), such that the flipping motion of a water monomer could be easily excited by the tunneling electrons during the IETS measurements, leading to time-averaged IETS of the initial and final states.

In order to accurately extract the vibrational energy from the IET spectra, especially from those with asymmetric line shapes, we fitted the IETS data using a model of resonantly enhanced IETS [23, 24]:

$$\sigma = \frac{m^2}{(\epsilon_a - \epsilon_F)^2 + (\Gamma/2)^2} \left\{ \frac{(\epsilon_a - \epsilon_F + \hbar\omega)^2 - \left(\frac{\Gamma}{2}\right)^2}{(\epsilon_a - \epsilon_F + \hbar\omega)^2 + \left(\frac{\Gamma}{2}\right)^2} \Theta(eV - \hbar\omega) \right.$$

$$\left. + \frac{\Gamma}{\pi} \frac{\epsilon_a - \epsilon_F + \hbar\omega}{(\epsilon_a - \epsilon_F + \hbar\omega)^2 + \left(\frac{\Gamma}{2}\right)^2} \ln\left|\frac{eV - \hbar\omega}{\hbar\omega}\right| \right\} \quad (6.1)$$

6.3 Impact of NQEs on the Strength of a Single H Bond

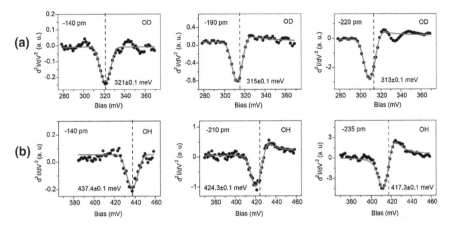

Fig. 6.5 IETS fitting for the H-bonded stretching modes of a HOD molecule at different tip heights. **a** OD stretching mode. **b** OH stretching mode. The tip height (indicated at the upper left of each graph) is referenced to the gap set with: V = 100 mV and I = 50 pA. The vibrational threshold energies obtained by IETS fitting are denoted by vertical dashed lines and presented in the panels. Reproduced with permission from Ref. [20]

Here, σ is the total tunneling conductance, m is the electron-vibration coupling constant, ϵ_F is the Fermi energy. ϵ_a, resonance state of the molecule, denotes the position of the molecular resonance with respect to E_F, and is coupled to the left and right electrode with constant Γ_L and Γ_R, respectively.

It should be noticed that the expression applies only when $\Gamma_{Tip} = \Gamma_L = \Gamma \gg \Gamma_R = \Gamma_{Sur}$. In our case, the water molecule is strongly coupled to the STM tip, and decoupled from the Au substrate by the NaCl bilayer film, thus satisfying the requirement. With this model, we achieve excellent agreement between simulated and experimental IETS data (Fig. 6.5).

6.3.2 NQEs of H-Bonding Strength

As we have discussed in the last section, OH and OD of the same water HOD monomer form H bonds with the Cl^- alternatively due to the flipping motion of water monomer on NaCl surface (Fig. 6.4b). This provides us a perfect opportunity to explore the impact of NQEs on the H-bonding strength by comparing the H1 and D1 modes in the same HOD molecule with the same STM tip. We plot the ratio between the vibrational frequency of H1 and D1 stretching modes as a function of tip-water distance (Fig. 6.6a). In general, as the STM tip approaches the water monomer, the ratio keeps on decreasing and crosses over the value of 1.361, which is the ratio of the free HOD monomer in gas phase, where without NQEs [25]. In addition to the general decreasing trend (region II), reversal behaviors appear at the large and small tip heights (region I and region III in Fig. 6.6a). The variation of the

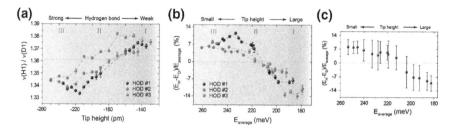

Fig. 6.6 Impact of NQEs on the H-bonding strength. a Tip height dependence of the ratio between the vibration frequency of H1 and D1 mode. The horizontal dashed line denotes the ratio (1.361) between the frequency of free OH and OD stretching modes. **b** Relative difference between the H-bonding energies of O–H⋯Cl (E_H) and O–D⋯Cl (E_D) as a function of their averaged energies ($E_{average} = (E_H + E_D)/2$). Different colors represent the results from three different HOD water monomers. **c** Same as (**b**), but is averaged over seven different HOD monomers. Reproduced with permission from Ref. [20]

ratio at different tip heights clearly indicates that the H-bonding strength changes upon isotope substitution because of the NQEs. Except for the similar trend, these curves obtained from different molecules differ from each other well beyond the experimental errors, which may arise from the variation of tip apex structure and the inhomogeneous local environment of the molecules resulting from the Au(111) substrate.

The H-bonding energy could be converted from the redshift of the H-bonded OH stretching frequency (relative to the free OH stretching energy) using an empirical formula [18]:

$$\Delta H = 1.3 \times \sqrt{\Delta \nu} \quad (6.2)$$

where ΔH is the H-bonding energy, in kJ/mol; $\Delta \nu$ is the redshift of the OH stretching mode, in cm^{-1}. We could also convert the unit of H-bonding strength to meV by: 1 kJ/mol = 10.4 meV/atom. To apply Eq. 6.2 to the OD stretching mode, the quantity $\Delta \nu$ should be multiplied by a factor: $\nu(OH)/\nu(OD) = 1.3612$ [25], where $\nu(OH)$ and $\nu(OD)$ are the OH and OD stretching frequencies of the free HOD molecule, respectively.

Using the empirical formula, we obtain the H-bonding energies from the redshifts of H1 and D1 vibrational frequency, which allow us to extract the impact of NQEs explicitly. Figure 6.6b is the relative difference of H-bonding energy between O–H⋯Cl and O–D⋯Cl as a function of their averaged energies. In general, the NQEs weaken the weak H bond and strengthen the relatively strong ones. But we noticed that the impact of the NQEs tends to diminish at the strong- and weak- H-bond limits (region I and region III in Fig. 6.6b). When taking the average of seven different groups of data (Fig. 6.6c), the crossover behavior of the NQEs is still observable, but the turning point at the strong- and weak-bond limits are smeared out. Therefore, the ability to probe the H-bonding strength at single bond limit is critical to accurately

6.3 Impact of NQEs on the Strength of a Single H Bond

assess the impact of NQEs. Surprisingly, the quantum component of the H bond can account for up to 14% of H-bonding strength, which is even bigger than the thermal energy at room temperature.

6.4 The Picture of Competing Quantum Effects

To explore the physical picture of the impact of NQEs on the H-bonding strength, we calculated the H1/D1···Cl distance and the corresponding H-bonding energy based on the PIMD simulations. As shown in Fig. 6.7a, b, the closer the tip is, the smaller H1/D1···Cl distance and the stronger H-bond is. The calculated NQEs and isotope effects on H-bonding interaction are both clearly demonstrated in Fig. 6.7c (black curve) by plotting the difference of H-bonding energy between H and D as a function of tip height. The simulation results nicely reproduce the experimental observations: the NQEs weaken the weak H bonds and strengthen the strong ones, and tend to fade out at the strong- and weak-bond limits.

Then, we compare the projection of O–H and O–D covalent bonds length along the intermolecular axis. Interestingly, when the tip approaches the water molecule, the ratio of the projections between O–H and O–D (Fig. 6.7c (red curve)) well reproduce

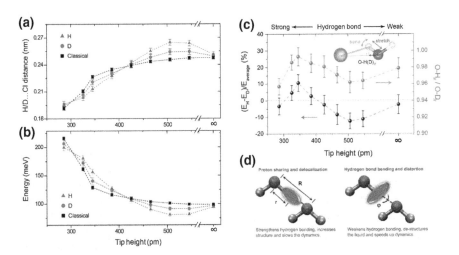

Fig. 6.7 Competing quantum effects in the H-bonding. a, b Averaged H1/D1···Cl distances **a** and binding energy $E_{H/D}$ **b** of the H-bond between the water and the NaCl surface at different tip heights. **c** Calculated H-bonding energy difference when replacing H with D (black) and the ratio between averaged projection of O–H and O–D covalent bonds along the intermolecular axis (red) at different tip height. Inset shows the projection geometry model. **d** Competing quantum effects in the H-bonding between two water molecules. There are two qualitatively different contributions to the vibrational ZPE, that is, O–H stretch (left) and bending (right) vibrational modes. **a–c** reproduced from Ref. [20] with permission, **d** from [26]

the behavior of the relative difference of the H-bonding energies between H and D (Fig. 6.7c (black curve)). Calculation results reveal that the anharmonic quantum fluctuations of O–H(D) mainly locate on the intramolecular covalent bond stretching and intermolecular H-bond bending (Fig. 6.7d). On the one hand, the zero-point motion of O–H(D) stretching increases the projection, resulting in the enhancement of the H-bonding strength (left panel in Fig. 6.7d). On the other hand, the zero-point motion of the H-bond bending makes the H-bond more bent, thus weakening the H bond (right panel in Fig. 6.7d). Therefore, it is the delicate competition between O–H(D) stretching and H-bond bending that induces the isotope effect on the H-bonding energy.

However, the reversal behaviors in regions I and III could not be predicted by this simple competing picture. For the turning point at the weak H-bond limit, it is easy to understand since the quantum contributions of the O–H(D) stretching and H-bond bending modes both diminish quickly and tend to cancel each other, resulting in the fade-out of the difference between the O–H⋯Cl and O–D⋯Cl bonding energies. At the strong-bond limit, the reversal feature is closely related to the noncolinear geometry of the O–H⋯Cl H-bond. Generally speaking, the stronger the H bond is, the more collinear the geometry becomes, but it is not true for O–H⋯Cl H-bond because of the repulsive interaction between the H^+ of the water and the Na^+ of NaCl surface, which leads to the noncolinear configuration of the O–H⋯Cl H bond. In Fig. 6.8a, we plot the force exerted on the H atom along the axis perpendicular to the O-H bond as a function of the tip height when the H atom is displaced toward the surface by 0.08 Å along this direction. We can see that as the tip moves down, the repulsive interaction between the H atom and the substrate increases. This is particularly true for the small tip height (<3.5 Å), i.e. strong H-bond case, which leads to the unusual non-collinear geometry of the strong H-bond.

In order to clarify how such a non-collinear geometry structure influences the NQEs, we plot the evolution of the potential energy surface (PES) at different tip heights as a function of the moving distance of the H atom along the H-bond bending motion (Fig. 6.8b). Note that the smaller tip height is, the stronger the H-bond gets. We use negative (positive) distance to denote the downward (upward) direction as denoted by the arrow (inset of Fig. 6.8b). It is clear that while the energy profiles on the right-hand side (distance > 0) change mildly as the tip approaches the surface, the change of these profiles on the left-hand side (distance < 0) becomes significantly steeper. The faster rise of the energy profile close to the surface means that the zero-point motion of H-bond bending mode tends to push the H atom further away from the surface, making the H-bond more bent. Therefore, at the small tip height (strong H-bond case) the anharmonic bending motion is greatly enhanced. This is in clear contrast to the case of normal collinear H-bond, where the anharmonic bending motion becomes less pronounced as the H-bond gets stronger. As already discussed in this section, it is the competition between covalent-bond stretching and H-bond bending that determines the behavior of NQEs (Fig. 6.7d). When the anharmonic quantum fluctuation of the bending mode dominates over that of the stretching mode at the small tip height, the reversal behavior of NQEs appears. This reversed behavior demonstrates that the NQEs of the H bond is extremely

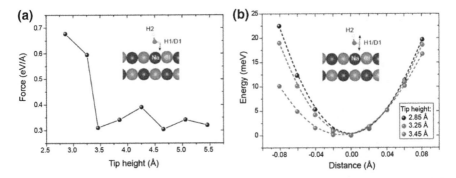

Fig. 6.8 Influence of the non-collinear geometry of the H-bond on the NQEs. **a** Forces H1 atoms feel when they were moved downward by 0.08 Å along the direction perpendicular to the O–H1 covalent bond and in the plane determined by the O, H1 and Na atom below the water molecule (denoted by an arrow in the inset). **b** Potential energy profiles along the axis corresponding to the H-bond bending motion (denoted by a double-ended arrow in the inset). Negative (positive) distance corresponds to the downward (upward) direction as denoted by the arrow (inset). Reproduced with permission from Ref. [20]

sensitive to the local environment, which is, at present, inaccessible by macroscopic spectroscopic methods.

We found that the picture of competing quantum effects exists in the recently discovered temperature dependence of the NQEs as well, in which the NQEs play a significant effect on the intermolecular H bond interaction in Watson-Crick CG base pair, from strengthening to weakening it with decreasing temperatures [27]. To explore the detailed mechanism, Fang et al. calculated the evolution of the occupation numbers of the bending and stretching modes at different temperatures as well as their impact on the intermolecular interactions, which demonstrated that the final influence of NQEs can be rationalized using the competition of different vibrational modes. At lower temperature, the bending modes can be highly populated compared with the stretching mode, leading to the weakening of the intermolecular H bond interaction. Therefore, in addition to the complexity induced by the noncolinear feature of the H bond, the picture of competition between the bending and stretching modes can be enriched by further considering the temperature effect. This competing quantum picture might be general and be applied to understand many abnormal properties of water [28].

6.5 Summary

In this chapter, isotopic substitution experiments combined with state-of-the-arts quantum simulations reveal that the anharmonic quantum fluctuations of hydrogen nuclei weaken the weak H bonds and strengthen the strong ones. However, this trend can be completely reversed when the H bond is strongly coupled to the polar atomic sites of the surface.

Those results unravel, for the first time, the quantum component of the H bond in a quantitative way, which turns out to be significantly larger than the thermal energy even at room temperature. In addition, this work highlights the important role of local environment in dictating the NQEs, which is at present inaccessible by conventional spectroscopic methods. It is particularly striking that the coupling to the atomic-scale species may completely reverse the widely accepted behavior of the NQEs, implying that the NQEs should be highly inhomogeneous and dynamic in nature. These findings substantially advance our understanding of the quantum nature of H bonds at the atomic level.

References

1. Benoit M, Marx D, Parrinello M (1998) Tunnelling and zero-point motion in high-pressure ice. Nature 392:258–261
2. Loubeyre P, LeToullec R, Wolanin E, Hanfland M, Husermann D (1999) Modulated phases and proton centring in ice observed by X-ray diffraction up to 170 GPa. Nature 397:503–506
3. Schwegler E, Sharma M, Gygi F, Galli G (2008) Melting of ice under pressure. Proc Natl Acad Sci USA 105:14779–14783
4. Morrone JA, Lin L, Car R (2009) Tunneling and delocalization effects in hydrogen bonded systems: a study in position and momentum space. J Chem Phys 130:204511
5. Li X-Z, Probert MIJ, Alavi A, Michaelides A (2010) Quantum nature of the proton in water-hydroxyl overlayers on metal surfaces. Phys Rev Lett 104:066102
6. Kumagai T et al (2010) Symmetric hydrogen bond in a water-hydroxyl complex on Cu(110). Phys Rev B 81:045402
7. Ubbelohde AR, Gallagher KJ (1955) Acid-base effects in hydrogen bonds in crystals. Acta Crystallogr 8:71–83
8. Matsushita E, Matsubara T (1982) Note on isotope effect in hydrogen bonded crystals. Prog Theor Phys 67:1–19
9. Nagata Y, Pool RE, Backus EHG, Bonn M (2012) Nuclear quantum effects affect bond orientation of water at the water-vapor interface. Phys Rev Lett 109:226101
10. Swalina C, Wang Q, Chakraborty A, Hammes-Schiffer S (2007) Analysis of nuclear quantum effects on hydrogen bonding. J Phys Chem A 111:2206–2212
11. Gregory JK, Clary DC (1996) Structure of water clusters. The contribution of many-body forces, monomer relaxation, and vibrational zero-point energy. J Phys Chem 100:18014–18022
12. Clary DC, Benoit DM, Van Mourik T (2000) H-densities: a new concept for hydrated molecules. Acc Chem Res 33:441–447
13. Voth GA, Chandler D, Miller WH (1989) Rigorous formulation of quantum transition state theory and its dynamical corrections. J Chem Phys 91:7749–7760
14. Tuckerman ME, Marx D, Klein ML, Parrinello M (1997) On the quantum nature of the shared proton in hydrogen bonds. Science 275:817–820
15. Morrone JA, Car R (2008) Nuclear quantum effects in water. Phys Rev Lett 101:017801
16. Paesani F, Voth GA (2009) The properties of water: insights from quantum simulations. J Phys Chem B 113:5702–5719
17. Li X-Z, Walker B, Michaelides A (2011) Quantum nature of the hydrogen bond. Proc Natl Acad Sci USA 108:6369–6373
18. Rozenberg M, Loewenschuss A, Marcus Y (2000) An empirical correlation between stretching vibration redshift and hydrogen bond length. Phys Chem Chem Phys 2:2699–2702

References

19. Guo J et al (2014) Real-space imaging of interfacial water with submolecular resolution. Nat Mater 13:184–189
20. Guo J et al (2016) Nuclear quantum effects of hydrogen bonds probed by tip-enhanced inelastic electron tunneling. Science 352:321–325
21. Ohara M, Kim Y, Yanagisawa S, Morikawa Y, Kawai M (2008) Role of molecular orbitals near the Fermi level in the excitation of vibrational modes of a single molecule at a scanning tunneling microscope junction. Phys Rev Lett 100:136104
22. Stiopkin IV et al (2011) Hydrogen bonding at the water surface revealed by isotopic dilution spectroscopy. Nature 474:192–195
23. Persson BNJ, Baratoff A (1987) Inelastic electron tunneling from a metal tip: the contribution from resonant processes. Phys Rev Lett 59:339–342
24. Baratoff A, Persson BNJ (1988) Theory of the local tunneling spectrum of a vibrating adsorbate. J Vac Sci Technol A 6:331–335
25. Iogansen AV (1999) Direct proportionality of the hydrogen bonding energy and the intensification of the stretching $v(XH)$ vibration in infrared spectra. Spectrochim Acta Part A 55:1585–1612
26. Ceriotti M et al (2016) Nuclear quantum effects in water and aqueous systems: experiment, theory, and current challenges. Chem Rev 116:7529–7550
27. Fang W et al (2016) Inverse temperature dependence of nuclear quantum effects in DNA base pairs. J Phys Chem Lett 7:2125–2131
28. Ceriotti M et al (2016) Nuclear quantum effects in water and aqueous systems: experiment, theory, and current challenges. Chem Rev 116:7529–7550

Chapter 7
Outlook

In this chapter I present an outlook on the directions of further studies of water-solid interfaces based on the STM/AFM technique, as well as the challenges in this research field. Technically, some new scanning probe techniques in combination with ultrafast technique and nitrogen-vacancy (NV) center are also introduced to overcome the intrinsic limitation of STM.

7.1 Perspective on Future Directions

7.1.1 Overlayer Ice

Although a majority of studies so far have been devoted to studying water overlayers deposited on hexagonal single crystal metal surfaces [1–11], there remains numerous issues under debate and questions to be unravelled. Feibelman reported a novel structure for water on Ru(0001) [12], that is, a partially dissociated layer to be energetically more favorable than the intact hexagonal bilayer ice model [13, 14]. The partial dissociation of water was further supported by x-ray photoelectron spectroscopy (XPS) [15] and STM studies, which showed the feature of hydroxyl groups. However, the partially dissociation was challenged by the vibrational spectroscopy study [16]. In addition, another XPS study performed by Andersson et al. [17] revealed that water wets the Ru(0001) surface nondissociatively at 150 K and energetic barrier of thermal dissociation is slightly larger than the thermal desorption. Wetting by water without dissociation has also been widely concluded for Pt(111) [5, 18–20]. But Lilach et al. found the dissociation of the first layer of water confined under the thick ice on Pt(111) [21]. Therefore, there is an urgent need to explore whether water dissociatively wets the surface or not, because dissociation of water plays an important role in electrochemistry, corrosion, environmental chemistry, or heterogeneous catalysis. What's more, it is also a key issue to understand the water dissociation mechanism, which is a crucial step in the development of efficient catalysts for splitting of water for hydrogen production. To probe this issue, it is highly vital to identify the detailed topography and O–H directionality of the H-bonded network using the STM and nc-AFM combined technique. It is also very interesting to explore the proton ordering, the edge structure and pre-melting of ice on solid surfaces.

7.1.2 Confined Water

It is well known that the confined water shows many exotic physical and chemical properties [22–27], which are rather distinct from those of the bulk water. Recently, water confined between the ultrathin 2D materials and various hydrophilic and hydrophobic substrates solid surfaces has attracted extensive attentions [28–31], and opens up the possibility of probing the water confined at solid/solid interfaces with STM. In particular, water confined between the graphene and mica surface forms ice-like structure at room temperature [28], but the detailed H-bonding structure is yet to be determined. Recently, Algara-Sille et al. reported the formation of a close packed square ice structure that confined between the hydrophobic graphene layers [30].

Another interesting system is water confined in the nanopores of the covalent organic framework (COF) and metal organic frameworks (MOF) on metal surfaces. The main advantage of the MOF and COF is that the size of the nanopores is adjustable by self-assembly technique. It would be interesting to study how the size of the lateral confinement affects the structure, dynamics as well as NQEs of water. It is also possible to achieve hydrophobic and hydrophilic confinements in a controlled manner by functionalizing the molecules with proper chemical groups. In addition to the lateral confinement, the vertical confinement can be implemented by approaching the STM tip to the water confined in the nanopores. The interaction between the water and the tip is also tunable by functionalizing the tip apex with different atoms or molecules [32–34]. In such a manner, the three dimensional (3D) cavity formed by the nanopores and the STM junction allows performing "high-pressure" experiments of water/ice with the scanning probe technique.

7.1.3 Water Hydration

Hydration at interfaces is of great importance for an extremely wide range of applied fields and processes, including salt dissolution [35], corrosion [36], electrochemistry [37], biological ion channel [38, 39], atmospheric aerosols [40] and water desalination [41]. One of the key issues is exploring the microscopic factors that govern the transport of the interfacial hydrated ions, especially for the nanoconfined fluidic systems [42–45]. In spite of massive experimental and theoretical efforts, the direct correlation between the structure and the transport properties of the hydrated ions is still lacking. The main difficulty lies in the interfacial inhomogeneity as well as the complex competing interaction among ions, water and surfaces, which requires detailed molecular-level characterization. Based on the submolecular-resolution imaging, tip manipulation and tip-enhanced IETS we developed with STM and nc-AFM, it is possible to further explore the hydration interaction between water molecules and other adsorbates on surfaces.

Salt surface provides an unprecedented opportunity for studying the ion hydration, because we have shown the possibility of pulling out individual cations and anions from the surface by the STM tip [32, 33, 46]. Thus, it is foreseeing that ion hydration clusters can be formed by artificially manipulating individual water molecules and ions on the salt surfaces. This capability is particularly useful to clarify the microscopic structure of the hydration shell and the mechanism of ionic-specific effects, which may provide new perspective to the initial stage of salt melting [47, 48], water desalination [41] and selective permeability of biological ion channel [39].

To investigate molecular hydration, it is desirable to choose the hydrophobic surfaces, such as graphene and Au, which on one hand minimize the interaction between the water and the surfaces, on the other hand enable the preparation of large molecules on the surfaces by spin-coating method or pulse injection method because of the inert nature of these substrate to the ambient environment. One noteworthy example of the molecular hydration is the protein hydration. SPM should be ideal in identifying the hydration sites and tracking the protein dynamics during the stepwise hydration process at single-molecule level by tip manipulation, which may shed new lights on the understanding of the competition between the water-water interaction and water-protein interaction.

7.1.4 Nuclear Quantum Effects

In spite of the advances we have achieved, the investigation of NQEs with STM is still in its infancy and there are many important issues yet to be solved. For instance, the studies in this thesis mainly focused on simple model systems. What if many-body correlation effects are entangled with the NQEs? It would be very interesting to study such correlated NQEs in larger water clusters or more extended H-bonded network such as overlayer ice, which might play key a role in the phase transition of ice. What's more, simulation results reveal that the concerted proton tunneling in hexagonal ice changes to thermally activated sequential over-barrier hopping at a surprisingly high transition temperature (~ 200 K) [49]. Hence, it is imperative to carry out NQEs experiments in a wide range of temperature to explore the transition temperature from quantum tunneling to classic hopping. As a matter of fact, besides hydrogen nuclei, NQEs might also exist in the materials or systems [50–52] that constructed by the other light elements (He/Li/N/C/B). It is very interesting to investigate how the isotope substitution influence the band structure and interlayer van der Waals interactions of 2D materials, such as BN and graphene, which might shed light on the new physics of the materials.

7.2 Challenges and New Possibilities

7.2.1 Ultrafast H-Bonding Dynamics

The bandwidth of the STM electronics is usually limited to MHz range, thus limits the application in probing the H-bonding dynamics in water such as proton transfer, bond-formation and breaking, and energy relaxation, which is typically happen at a rather short time scale within picosecond (ps) or even femtosecond (fs) range. Ultrafast technique combined STM has been proven to be an efficient tool to overcome this limitation and could achieve both Ångström spatial resolution and femtosecond temporal resolution simultaneously [53, 54]. In such a setup, two laser pulses with a time delay are focused into the STM junction, exciting the molecules on the surface one after the other. The excitation of the molecule will induce a transient change of the tunneling current. If the molecule excited by the first pulse does not relax to the ground state at the moment when the second pulse arrives, the second pulse cannot induce a current change. Changing the delay between the two laser pulses thus leads to a variation of the average tunneling current, which could be readily measured by the slow STM electronics. Therefore, the temporal resolution is only limited by the width of the laser pulse.

The application of laser-combined STM has been successfully demonstrated on semiconductor surfaces, focusing on the charge carrier dynamics and spin relaxation dynamics [53, 54]. In addition to charge and spin dynamics, the optical pump-probe STM can be also used to probe the nuclear dynamics. Very recently, Li et al. investigated the coherent vibration driven conformational changes of a single molecule using the femtosecond laser-combined STM [55]. The laser-combined STM is particularly useful to study the H-bonding dynamics of interfacial and confined water, which are susceptible to the coupling with the local environments. It is foreseeing that the laser-STM will become a powerful tool in the near future for investigating various ultrafast dynamics processes at the single molecule level, which may reshape our current knowledge of water/solid interfaces.

7.2.2 Non-invasive Measurement of Nuclear Spin of Proton

Another intrinsic problem of scanning probe microscopes (SPM) is that all the probes inevitably induce perturbation to the water structure, due to the excitation of the tunneling electrons and the tip-water interacting forces. In addition, high-resolution STM/AFM measurements usually require ultrahigh vacuum (UHV) conditions and low temperatures, otherwise the signal to noise ratio of imaging and spectroscopy would be very poor and atomic-scale investigation would become difficult due to the unstable of the tip apex. Those limitations make SPM fall short compared with the conventional spectroscopic methods such as optical spectroscopy, neutron scattering and nuclear magnetic resonance (NMR). Recently, an emerging SPM technique, which employs the nitrogen-vacancy (NV) center as the scanning probe (NV-SPM), shows great potential to overcome those limitations, and may become an ideal non-perturbative tool for the study of interfacial water systems under ambient conditions.

NV defect centers naturally existing in diamond serve as nanoscale magnetometers because of its atomic size and proximity to diamond surface (<10 nm). Its lone-pair electrons result in a triplet ground state, whose spin state can be polarized by laser excitation and readout by spin-dependent fluorescence. Long-coherence time (0.1–1 ms) [56] makes such solid-state quantum sensors stable at ambient conditions and easy to be coherently manipulated by microwave pulse sequence. Unknown target spins and magnetic field could be sensed by a single NV center through magnetic dipole interaction, which allows the detection of weak signals such as spin fluctuations of protons in water at a probe-sample distance of 5–20 nm [57, 58]. Meanwhile, the power of excitation/readout laser is as low as several tens of microwatt, so the laser-heating effect is negligible. Therefore, NV-SPM is almost non-invasive to the water structure during the imaging.

What's more, NV-SPM is able to conduct NMR spectroscopy at nanoscale or even single molecule level. Similar to RF pulse-based technology ubiquitously applied in conventional NMR experiments, modified elaborate microwave pulse sequence endows NV-SPM a variety of spectroscopic technologies with a high resolution (~10 kHz) and a large bandwidth (DC to ~3 GHz) [59]. For example, intramolecular proton magnetic resonance of a single water molecule could be resolved by an extremely proximal NV center or by the quantum state transference through reporter spin near the target molecule [60], proton correlations [61] and molecular dynamics [62] could be analyzed from relaxation/dephasing spectroscopy mapping with nanoscale spatial resolution.

References

1. Morgenstern K (2002) Scanning tunnelling microscopy investigation of water in submonolayer coverage on Ag(111). Surf Sci 504:293–300
2. Cerda J et al (2004) Novel water overlayer growth on Pd(111) characterized with scanning tunneling microscopy and density functional theory. Phys Rev Lett 93:116101
3. Dulub O, Meyer B, Diebold U (2005) Observation of the dynamical change in a water monolayer adsorbed on a ZnO surface. Phys Rev Lett 95:136101
4. Feibelman PJ (2010) The first wetting layer on a solid. Phys Today 63:34–39
5. Nie S, Feibelman PJ, Bartelt NC, Thuermer K (2010) Pentagons and heptagons in the first water layer on Pt(111). Phys Rev Lett 105:026102
6. Forster M, Raval R, Hodgson A, Carrasco J, Michaelides A (2011) c(2 x 2) Water-hydroxyl layer on Cu(110): a wetting layer stabilized by Bjerrum defects. Phys Rev Lett 106:046103
7. Chen J et al (2014) An unconventional bilayer ice structure on a NaCl(001) film. Nat Commun 5:4056
8. Thurmer K, Nie S, Feibelman PJ, Bartelt NC (2014) Clusters, molecular layers, and 3D crystals of water on Ni(111). J Chem Phys 141:18C520
9. Maier S, Lechner BAJ, Somorjai GA, Salmeron M (2016) Growth and structure of the first layers of ice on Ru(0001) and Pt(111). J Am Chem Soc 138:3145–3151

10. Haq S, Clay C, Darling GR, Zimbitas G, Hodgson A (2006) Growth of intact water ice on Ru(0001) between 140 and 160 K: Experiment and density-functional theory calculations. Phys Rev B 73:115414
11. Hodgson A, Haq S (2009) Water adsorption and the wetting of metal surfaces. Surf Sci Rep 64:381–451
12. Feibelman PJ (2002) Partial dissociation of water on Ru(0001). Science 295:99–102
13. Thiel PA, Madey TE (1987) The interaction of water with solid surfaces: fundamental aspects. Surf Sci Rep 7:211–385
14. Henderson MA (2002) The interaction of water with solid surfaces: fundamental aspects revisited. Surf Sci Rep 46:1–308
15. Weissenrieder J, Mikkelsen A, Andersen JN, Feibelman PJ, Held G (2004) Experimental evidence for a partially dissociated water bilayer on Ru{0001}. Phys Rev Lett 93:196102
16. Denzler DN et al (2003) Interfacial structure of water on Ru(001) investigated by vibrational spectroscopy. Chem Phys Lett 376:618–624
17. Andersson K, Nikitin A, Pettersson LGM, Nilsson A, Ogasawara H (2004) Water dissociation on Ru(001): an activated process. Phys Rev Lett 93:196101
18. Ogasawara H et al (2002) Structure and bonding of water on Pt(111). Phys Rev Lett 89:276102
19. Clay C, Haq S, Hodgson A (2004) Hydrogen bonding in mixed $OH+H_2O$ overlayers on Pt(111). Phys Rev Lett 92:046102
20. Verdaguer A, Sacha GM, Bluhm H, Salmeron M (2006) Molecular structure of water at interfaces: Wetting at the nanometer scale. Chem Rev 106:1478–1510
21. Lilach Y, Iedema MJ, Cowin JP (2007) Dissociation of water buried under ice on Pt(111). Phys Rev Lett 98:016105
22. Reiter G et al (2006) Anomalous behavior of proton zero point motion in water confined in carbon nanotubes. Phys Rev Lett 97:247801
23. Garbuio V et al (2007) Proton quantum coherence observed in water confined in silica nanopores. J Chem Phys 127:154501
24. Reiter GF et al (2012) Evidence for an anomalous quantum state of protons in nanoconfined water. Phys Rev B 85:045403
25. Agrawal KV, Shimizu S, Drahushuk LW, Kilcoyne D, Strano MS (2017) Observation of extreme phase transition temperatures of water confined inside isolated carbon nanotubes. Nat Nanotech 12:267–273
26. Dellago C, Naor MM, Hummer G (2003) Proton transport through water-filled carbon nanotubes. Phys Rev Lett 90:105902
27. Kolesnikov AI et al (2004) Anomalously soft dynamics of water in a nanotube: a revelation of nanoscale confinement. Phys Rev Lett 93:035503
28. Xu K, Cao P, Heath JR (2010) Graphene visualizes the first water adlayers on mica at ambient conditions. Science 329:1188–1191
29. Cao P, Xu K, Varghese JO, Heath JR (2011) The microscopic structure of adsorbed water on hydrophobic surfaces under ambient conditions. Nano Lett 11:5581–5586
30. Algara-Siller G et al (2015) Square ice in graphene nanocapillaries. Nature 519:443–445
31. Feng XF, Maier S, Salmeron M (2012) Water splits epitaxial graphene and intercalates. J Am Chem Soc 134:5662–5668
32. Meng X et al (2015) Direct visualization of concerted proton tunnelling in a water nanocluster. Nat Phys 11:235–239
33. Guo J et al (2016) Nuclear quantum effects of hydrogen bonds probed by tip-enhanced inelastic electron tunneling. Science 352:321–325
34. Peng JB et al (2018) Weakly perturbative imaging of interfacial water with submolecular resolution by atomic force microscopy. Nat Commun 9:112
35. Klimes J, Bowler DR, Michaelides A (2013) Understanding the role of ions and water molecules in the NaCl dissolution process. J Chem Phys 139:234702
36. Bunker BC (1994) Molecular mechanisms for corrosion of silica and silicate glasses. J Non-Cryst Solids 179:300–308

37. Armand M, Endres F, MacFarlane DR, Ohno H, Scrosati B (2009) Ionic-liquid materials for the electrochemical challenges of the future. Nature Mater. 8:621–629
38. Gouaux E, MacKinnon R (2005) Principles of selective ion transport in channels and pumps. Science 310:1461–1465
39. Payandeh J, Scheuer T, Zheng N, Catterall WA (2011) The crystal structure of a voltage-gated sodium channel. Nature 475:353–358
40. Sipila M et al (2016) Molecular-scale evidence of aerosol particle formation via sequential addition of HIO_3. Nature 537:532–534
41. Cohen-Tanugi D, Grossman JC (2012) Water desalination across nanoporous graphene. Nano Lett 12:3602–3608
42. Maffeo C, Bhattacharya S, Yoo J, Wells D, Aksimentiev A (2012) Modeling and simulation of ion channels. Chem Rev 112:6250–6284
43. Guo W, Tian Y, Jiang L (2013) Asymmetric ion transport through ion-channel-mimetic solid-state nanopores. Acc Chem Res 46:2834–2846
44. Jain T et al (2015) Heterogeneous sub-continuum ionic transport in statistically isolated graphene nanopores. Nat Nanotech 10:1053
45. Tagliazucchi M, Szleifer I (2015) Transport mechanisms in nanopores and nanochannels: can we mimic nature? Mater. Today 18:131–142
46. Peng J, Guo J, Ma R, Meng X, Jiang Y (2017) Atomic-scale imaging of the dissolution of NaCl islands by water at low temperature. J Phys: Condens Matter 29:104001
47. Rossi MJ (2003) Heterogeneous reactions on salts. Chem Rev 103:4823–4882
48. Liu L-M, Laio A, Michaelides A (2011) Initial stages of salt crystal dissolution determined with ab initio molecular dynamics. Phys Chem Chem Phys 13:13162–13166
49. Drechsel-Grau C, Marx D (2017) Collective proton transfer in ordinary ice: local environments, temperature dependence and deuteration effects. Phys Chem Chem Phys 19:2623–2635
50. Cardona M, Thewalt MLW (2005) Isotope effects on the optical spectra of semiconductors. Rev Mod Phys 77:1173–1224
51. Vuong TQP et al (2018) Isotope engineering of van der Waals interactions in hexagonal boron nitride. Nat Mater 17:152
52. Patrick CE, Giustino F (2013) Quantum nuclear dynamics in the photophysics of diamondoids. Nat Commun 4:2006
53. Terada Y, Yoshida S, Takeuchi O, Shigekawa H (2010) Real-space imaging of transient carrier dynamics by nanoscale pump-probe microscopy. Nat Photon 4:869–874
54. Yoshida S et al (2014) Probing ultrafast spin dynamics with optical pump-probe scanning tunnelling microscopy. Nat Nanotech 9:588–593
55. Li SW, Chen SY, Li J, Wu RQ, Ho W (2017) Joint space-time coherent vibration driven conformational transitions in a single molecule. Phys Rev Lett 119:176002
56. Maze JR et al (2008) Nanoscale magnetic sensing with an individual electronic spin in diamond. Nature 455:644-U641
57. Mamin HJ et al (2013) Nanoscale nuclear magnetic resonance with a nitrogen-vacancy spin sensor. Science 339:557–560
58. Staudacher T et al (2013) Nuclear magnetic resonance spectroscopy on a (5-nanometer)^3 sample volume. Science 339:561–563
59. Tetienne JP et al (2016) Scanning nanospin ensemble microscope for nanoscale magnetic and thermal imaging. Nano Lett 16:326–333
60. Shi FZ et al (2015) Single-protein spin resonance spectroscopy under ambient conditions. Science 347:1135–1138
61. Laraoui A et al (2013) High-resolution correlation spectroscopy of ^{13}C spins near a nitrogen-vacancy centre in diamond. Nat Commun 4:1651
62. Staudacher T et al (2015) Probing molecular dynamics at the nanoscale via an individual paramagnetic centre. Nat Commun 6:8527